THE CHEMISTRY OF METALLIDES

METALLIDY I VZAIMODEISTVIE MEZHDU NIMI

МЕТАЛЛИДЫ И ВЗАИМОДЕЙСТВИЕ МЕЖДУ НИМИ

THE CHEMISTRY OF METALLIDES

Ivan Ivanovich Kornilov
Director, Metallic Alloy Chemistry Laboratory
A. A. Baikov Metallurgy Institute
Academy of Sciences of the USSR, Moscow

Translated from Russian by
J. W. Loweberg

Springer Science+Business Media, LLC
1966

Ivan Ivanovich Kornilov, Director of the Metallic Alloy Chemistry Laboratory, A. A. Baikov Metallurgy Institute of the Academy of Sciences of the USSR, was born in 1904. From 1930 to 1934 he worked at the Academy of Sciences of the USSR under Academician N. S. Kurnakov investigating phase equilibria and transformations in metallic systems. Afterwards he worked at the Institute of General and Inorganic Chemistry of the Academy of Sciences of the USSR continuing his research on equilibrium diagrams of metallic systems and the properties of alloys. In 1937 he was appointed director of the Metallic Alloy Chemistry Laboratory at the institute.

In 1945 Kornilov published *Iron Alloys, Vol. I: Alloys of the Iron-Chromium-Aluminum System* based on his systematic research on the equilibrium diagram of the ternary system iron-chromium-aluminum and the development of new high-temperature oxidation-resistant alloys. Two additional volumes in this series by Kornilov appeared subsequently: *Iron Alloys, Vol. II: Solid Solutions of Iron* (1951) and *Iron Alloys, Vol. III: Iron-Nickel-Chromium Alloys* (1956). In 1958 Kornilov published *Nickel and Its Alloys, Vol. I: Nickel*. Kornilov's *Physicochemical Bases of Heat Resistance in Alloys*, concerning equilibrium diagrams of multicomponent metallic systems and the relation between equilibrium diagrams and the heat resistance of alloys, appeared in 1962. Kornilov and his co-workers have published about 250 scientific articles including several in English, German, and French. Kornilov is also Assistant Director of the Science Department at the Metallurgy Institute.

The Russian text, originally published for the State Committee for Ferrous and Nonferrous Metallurgy of the State Planning Commission of the USSR and the A. A. Baikov Metallurgy Institute by Nauka Press in Moscow in 1964, has been extensively revised and enlarged for this edition by the author.

Металлиды и взаимодействие между ними
Иван Иванович Корнилов

Library of Congress Catalog Card Number 65-23386

© *1966 Springer Science+Business Media New York*
Originally published by Consultants Bureau in 1966.
Softcover reprint of the hardcover 1st edition 1966

ISBN 978-1-4899-4748-2   ISBN 978-1-4899-4746-8 (eBook)
DOI 10.1007/978-1-4899-4746-8

FOREWORD

Metallides are formed by reactions between metals or metals and nonmetals. They differ from other compounds in the nature of their chemical bonding. Their diverse compositions and structures determine their physical properties, which are different from the properties of pure metals and alloys. Owing to these properties they are widely used to develop new high-strength, heat-resistant, and chemically stable materials. They include compounds having special superconductor, semiconductor, magnetic, optical, and other properties. Metallides of various classes have found especially wide use recently in connection with the general progress of new branches of engineering (atomic energy, aviation engineering, radio engineering, electronics, etc.).

The properties of individual compounds vary regularly with the character of their reactions among themselves. Hence investigation of reactions between metallides is interesting in itself. In one section of the book the author develops ideas which he proposed earlier on reactions between metallides, i.e., formation of solid solutions, compounds, and mechanical mixtures in binary, ternary, and more complex metallide systems. These questions comprise one of the main divisions of metal chemistry.

The present monograph, while making no claim to exhaustive completeness, characterizes adequately the current state of the theory of metallide reactions. The author indicates the main factors determining the conditions of formation of solid solutions and heterogeneous alloys based on metallides and presents fields of application of metallides as a new class of inorganic materials.

PREFACE

Questions of the chemical interaction of metals and elements of the periodic system occupy an important place in the development of modern inorganic chemistry. The main problems of this branch of science, which is called metal chemistry, include questions of the formation of metallic liquid and solid solutions, compounds of metals, and solid-state transformation reactions.

Numerous investigations of separate and specific questions of metal chemistry have been conducted, the results of which are scattered throughout many journal articles and transactions of scientific conferences. These problems were discussed in many symposia on various subjects: theory of metal alloys, in Moscow (1953); physical chemistry of metallic solutions and compounds, in Philadelphia (1959); theory of solid solutions, in Kiev (1960); nature of metallic solid solutions, in Paris (1961); electronic structure and chemistry of transition metal alloys, in New York (1962).

A number of collections and separate monographs on solid solutions and compounds of metals were published as a result of the discussion of the papers and reports in these scientific symposia. We should also take note of recently published scientific papers on separate classes of inorganic compounds: refractory compounds of metals, semiconductor compounds, and superconductor materials, including superconductor compounds.

Until now, however, there has been no special publication devoted to the formation of compounds of metals, establishment of the relation between the position of metals and nonmetals in the periodic system and the composition and chemical structure of metal compounds, and reactions between different types of compounds.

As it seems to us, study of the factors and mechanism of formation of compounds of metals with metals and nonmetals, classification of various types of such compounds, and the problem of chemical bonding and changes in properties with respect to composition and structure for these compounds are interesting in themselves in the overall development of investigations of inorganic materials.

It is especially important to consider these questions of metal chemistry because major contradictions in the understanding of the physicochemical nature of metal compounds have always existed in this field; there is no single system of classification and terminology for such compounds, and their interactions have received little study. In an early stage of development of alloy chemistry, compounds of metals with metals acquired the name "metallic compounds" (N. S. Kurnakov), or "metallides" (N. S. Kurnakov and N. I. Stepanov); other terms were used later in courses on metallurgy and metal physics: "intermetallic compounds," "intermetallides," "intermediate phases," "ordered structures," etc. As a matter of practical terminology, the concept "intermetallic compounds" necessarily excluded many compounds of metals and nonmetals, even compounds having metallic-type chemical bonding, such as, borides, carbides, silicides, and nitrides. There are many compounds with covalent bonding which also had to be excluded from this group. Most of the latter are now included in a separate group of semiconductor compounds. Moreover, opinion was divided on the terminology of compounds of metals with phosphorus, sulfur, and oxygen, which react to form a wide variety of compounds having different types of chemical bonding.

Depending on composition, many metal compounds have metallic, covalent, or ionic bonding or various combinations of intermediate, mixed types of bonding. As a rule, the compounds listed above do not conform to valence theory, and the atomic ratios in them are determined by the character of electron distribution in the atoms of their crystal lattices and the type of chemical bonding which appears in this case. Only compounds of metals with halogen-group (VII) elements have purely ionic bonding and are regarded as typical inorganic, saltlike compounds with atomic ratios related to valence. Thus in the case of compounds of metals with other elements, one can follow their varied chemical character and mutual transitions of different types of bonding from metallic to ionic.

The formation of metal compounds other than those with ionic bonding is treated in this book. We considered it expedient to group all such compounds under the general term "metallides," which, as it seems to us, encompasses a wide variety of metal compounds in the most general form and from a single chemical viewpoint.

The following principal physicochemical properties are adopted in this book as the main factors determining the conditions of formation of metallides and their interactions: the electronic structures of the atoms, ratios of atomic radii, electronegativities, and ionization potentials of the elements with respect to their relative position in the periodic system.

Based on similarities and differences in these properties, which we call the metallochemical properties of the elements, a classification of metallides having different types of chemical bonding is given, and the possibilities of mutual transitions of type of bonding from metallic to ionic and the presence of intermediate and mixed types of bonding in various metal compounds are shown. The broad range of variation in the character of chemical bonding among metallides determines their varied chemical, physical, and mechanical properties, which differ from the properties of pure metals, solid solutions, and metallic alloys. Owing to their specific properties, metallides have been used widely in the development of new materials having high hardness, heat resistance, and chemical stability and special physical properties. Among them there are compounds having unique superconductor, semiconductor, magnetic, optical, and other properties. Recently the use of metallides of various classes has increased markedly, especially in connection with the overall progress of new branches of engineering — atomic power, rocket and aviation engineering, radio engineering, electronics, etc. One may confidently state that many designs of modern and new machines and devices used in atomic engineering, radio engineering, and electronics, new superpowerful magnets, and high-efficiency quantum generators could not be realized without developing new magnetic, semiconductive, superconductive, and other metal compounds.

These remarkable properties of individual compounds vary regularly with respect to the character of their interactions, i.e., the formation among them of solid solutions and compounds having more complex chemical compositions and types of crystal structure than simple binary compounds. These properties vary within broad limits, depending on the different metastable and stable states of metallide systems.

Hence, it is very interesting to investigate reactions between metallides and to study the resulting solutions, complex compounds, and alloys based on them. Some rules of formation of solid solutions among metallides, which we established earlier (1951), were confirmed in further investigations by many authors. The theoretical and experimental materials on metallide reactions, considered in this book, were based on them. This book presents the development of concepts of reactions between metallides, i.e., the formation of solid solutions, many compounds, and mechanical mixtures in binary, ternary, and many-component metallide systems. The rules of change in properties with composition in such metallide systems conform to Kurnakov's fundamental laws, which he established in the field of physicochemical analysis. In degree, however, these properties differ markedly from the properties of simple metallic systems.

Although the present monograph does not claim to be a complete review of the literature, the current state of the theory of metallide interactions is adequately characterized in it, and the main factors determining the conditions of formation of solid solutions and heterogeneous alloys based on metallides are described, as well as many fields of application of metallides and alloys based on them.

Rules applying to the field of metallide chemistry, established in this monograph, are discussed in its concluding part. The author uses them to predict the existence of many metallide systems in which one should expect the formation of solid solutions and heterogeneous structures. Further investigations along the lines set forth above in the field of metal chemistry are important for working out the general theory of solids and the physicochemical basis for developing new inorganic materials with special physical properties.

April 20, 1964   I. I. Kornilov

CONTENTS

PUBLISHER'S NOTE

The following Soviet journals cited in this book are available in cover-to-cover translations:

Russian Title	English Title	Publisher
Doklady Akademii Nauk SSSR	Doklady Chemistry	Consultants Bureau
Fizika metallov i metallovedenie	Physics of Metals and Metallography	Acta Metallurgica
Izvestiya Akademii Nauk SSSR: Seriya khimicheskaya	Bulletin of the Academy of Sciences of the USSR: Division of Chemical Science	Consultants Bureau
Uspekhi khimii	Russian Chemical Reviews	Chemical Society (London)
Zhurnal éksperimental'noi teoreticheskoi fiziki	Soviet Physics—JETP	American Institute of Physics
Zhurnal fizicheskoi khimii	Russian Journal of Physical Chemistry	Chemical Society (London)
Zhurnal neorganicheskoi khimii	Russian Journal of Inorganic Chemistry	Chemical Society (London)
Zhurnal strukturnoi khimii	Journal of Structural Chemistry	Consultants Bureau

CHAPTER I

METAL CHEMISTRY AND METALLIDES

History of the Subject

Although metals and alloys played an important part in the development of human society and culture even in ancient times, they did not receive attention as objects of chemical investigations until the 17th and 18th centuries. In M. V. Lomonosov's first papers on physics, chemistry, and metallurgy [1, 2] it is stated that metals combine when alloyed. He considered the formation of alloys of different metals from the viewpoint of interaction of metals in solutions.

Statements regarding the chemical theory of formation of metallic alloys appear in papers by chemists of the early 19th century [3, 4]. Hypotheses regarding the existence of chemical compounds between metals were first advanced in [5], which appeared in the 1840's. Here we find the first reports of investigations in the field of chemical methods for isolating iron carbide from steels and establishing its formula, Fe_3C. These publications were the basis for development of the so-called method of phase analysis, which now plays an important part in determining the compositions of various phases in complex metal alloys. Matthiessen [6] drew important conclusions regarding the chemical nature of metal alloys, based on physical methods of investigating alloys with respect to composition.

It must be admitted, however, that systematic investigations of combinations of metals did not begin until the end of the 19th century [7]. Mendeleev devoted considerable attention to such questions [7]. He considered metal alloys from the viewpoint of his classical opinions on solutions and unsaturated compounds. In "Fundamentals of Chemistry" it is stated directly that "other unsaturated chemical compounds, e.g., metal alloys, are similar to solutions" (p. 440). Noting the importance of investigating metal alloys, Mendeleev writes: "Today solutions and alloys are usually studied, since I believe that the study of solutions and especially alloys, which at present is still in an embryonic state, will in time clear up a good deal with regard to chemical forces and phenomena" (pp. 581-582).

In studying the general character of chemical reactions between metals, Mendeleev's periodic law plays the main part. The principal rules of interaction of metals and formation of metallic solutions and compounds are based on this law. A striking example of application of the periodic system to the study of intermetallic compounds is a paper by N. S. Kurnakov, published in 1899 [8]. In it (p. 5) compounds formed by the combination of metals are discussed from the chemical viewpoint for the first time, and attention is directed to the tendency of alkali and alkaline-earth metals, as the chief bearers of metallic properties, to react in this way. He gave such compounds of metals the name of metallic (or positive) compounds. The question of the nomenclature of metal compounds will be considered below.

The papers of W. Gibbs [9] played an important part in developing investigations of equilibria in chemical systems, including metallic ones. His papers on heterogeneous equilibria and the phase rule proposed by him led, in Kurnakov's words, to the creation of an international language based on the geometrical form of equilibrium-system diagrams.

Research on metallic alloys since the end of the 19th century also was developed in these theoretical directions abroad by Tammann in Germany [10], Austen in England [11], Le Chatelier in France [12], among others, Chemical reactions between metals were considered in a special monograph [13], which today has become a bibliographic rarity. Subsquent investigations of reactions between metals, equilibrium diagrams, and phase transitions were conducted in connection with the general progress of physicochemical analysis [14, 15]; metallurgy [16-19], and metal physics [20-23].

Definition of Metal Chemistry

   The chemical trend in the study of metal alloys, initiated by Kurnakov on the basis of applying the periodic law and studying equilibria, was called metallic-alloy chemistry [24, 25] and in recent years has been grouped under the general term metal chemistry [26-28]. It is based on a combination of inorganic and physical chemistry and related to metallurgy, crystal chemistry, and solid-state physics and chemistry.

   The general problems of metal chemistry, determined in the course of its development, were formulated in a report [28], delivered to the General Conference of the Division of Chemical Sciences, Academy of Sciences, USSR in 1956, in the following manner: "Study of the chemical interaction of metallic elements of the periodic system with one another or with nonmetals in the case where solutions and compounds with metallic bonding are formed by the latter, and establishment of general rules of formation, structure, and properties of a broad class of metal solutions and compounds." As a rule, such metal compounds do not conform to the theory of valence. This characteristic difference in the chemistry of metallic elements, the formation of many metal compounds which do not conform to valence theory, makes it richer and more complex for investigation than the chemistry of the simplest inorganic compounds, where atomic ratios are governed by valence theory.

   There is no class of inorganic substances, to say nothing of organic ones, in which the capacity for isomorphous displacement or introduction of foreign atoms into the crystal lattice is as strongly manifested as in formation of various metallic solid solutions. The latter are formed also between metallic compounds. In this respect, solid solutions of metals and metallic compounds can be compared to some degree only with natural inorganic compounds — minerals, where the capacity for isomorphous displacement of a number of compounds (e.g., oxides, silicates, spinels, etc.) also is manifested to a considerable degree, and iron meteorites, which consist mainly of solid solutions of iron, nickel, and cobalt.

   Metal chemistry encompasses the chemical interactions of more than 80 elements of the periodic system which are metals. It should be added that many reaction products of metals and nonmetals (boron, carbon, silicon, nitrogen, and even phosphorus, sulfur, and oxygen) have metallic bonding. Solutions and compounds of metals with these nonmetals are considered in metal chemistry. In many respects such compounds are transitional links between purely metallic compounds and compounds having covalent and ionic bonding.

   The province of metal chemistry is the study of:

   1) the main factors determining the interactions of metals;
   2) the formation of metal solid solutions;
   3) the formation of metallic compounds;
   4) crystallochemical reactions in metallic systems;
   5) reactions between metallic compounds.

   The items listed above have been and are being extensively investigated by chemists, physicists, and metallurgists. During the last 50 years many papers have been published in this field and substantial results achieved. Many current problems of metal chemistry were discussed in a special symposium on the physical chemistry of metallic solutions and compounds, held in London [9]. In the transactions of this symposium the main physicochemical factors determining the formation of solid solutions in some systems and compounds in others were considered, as well as modern methods for investigating metal alloys.

Main Factors Determining the Interactions of Metals

   Despite the apparent simplicity in the interrelations of metals, the conditions under which complex substances with metallic bonding are formed cannot yet be regarded as definitely established. Questions of electron distribution and chemical bonding in complex metallic solutions and compounds are especially difficult to explain. At present there are certain generally recognized physicochemical factors [16-29] which, if not in full measure, at least to a considerable degree determine the conditions of formation of metallic solid solutions and metallic compounds.

The most important condition for characterizing reactions between metals is their relative position in the periodic system. This determines the chemical properties of the elements, as well as the similarity and difference in electronic structure of the atoms.

Generalizing the opinions expressed in the literature by various authors, one must recognize that the most important factors determining the character of interaction of metals among themselves and with nonmetals are the following: 1) the atomic radii of the metals; 2) the positions of the metals in the electronegativity series of elements; 3) the valences and ionization potentials of the elements.

All these properties of the elements depend on their position in the periodic system. In application to interactions of metals they may be called the metallochemical properties of the elements. We shall use this name extensively in this book. None of these properties is sufficient in itself for studying reactions between metals and nonmetals. They must be considered in combination to draw correct conclusions regarding the character of interaction of metal solutions and compounds and the types of chemical bonding which appear in this case. The effect of various factors on the formation of metallic solutions and compounds is considered briefly below.

Atomic Radii of Metals

Many reactions of metals with nonmetals or between metals differ from ordinary chemical reactions (ionic, molecular, etc.) by their atomic character and the formation of crystalline solutions and compounds with characteristic types of metallic and covalent bonding between atoms. Owing to these peculiarities the interatomic distances, determined by the atomic radii of the elements, are important.

Atomic radii, like many other properties of elements, are strictly periodic functions of atomic number. The atomic radii of the elements are plotted in Fig. 1 from modern data. Contrary to earlier methods of representing them by periods, they are shown here by groups of elements in the periodic system. In this figure beryllium, boron, and aluminum are located in Groups II and III, subgroup B, where they show more regularity in change of atomic radius than in subgroup A.

With respect to atomic radius, all elements may be divided into three series. One of them comprises elements having intermediate atomic radii, whereas the two outside series comprise elements with the largest and smallest atomic radii.

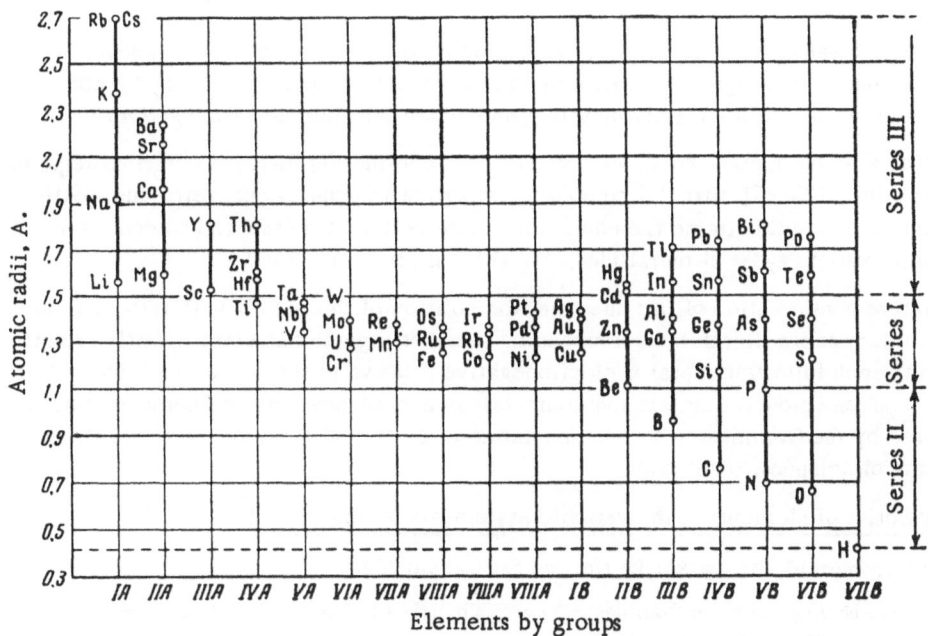

Fig. 1. Atomic radii of elements by groups of the periodic system.

The following characteristics may be noted in Fig. 1: Metals of Groups IA and IIA having small atomic numbers, which are the most typical metals, have the largest atomic radii; elements of Groups IIIB—VIIB, which are typical light nonmetals (B, C, N, O, H) have the smallest. The atomic radii of these elements decrease in the given order from B to H. In the case of heavy metals of Groups IIIB—VB having large atomic numbers (In, Tl, Sn, Pb, Sb, Bi) the atomic radii increase with atomic number. Metals which are transition elements (Groups III—VIII) with unfilled d-electron shells, as well as metals of Groups IB and IIB, have intermediate atomic radii and are included in the first series.

Within a given group the atomic radius increases with atomic number. Elements which are very close analogs, such as Ti, Zr, and Hf; V, Nb, and Ta; Cr, Mo, and W; Fe, Co, and Ni; Ag and Au, etc., have very nearly the same atomic radii. However, some pairs of analogous elements have considerably different atomic radii, e.g., metals of Groups IA and IIA, Al and B, Si and C, etc.

The similarity or difference in atomic radius of elements, together with other properties of atoms, is of decisive importance in interactions of metals, particularly when compounds are formed. The relative position of elements in subgroups A and B is especially important in this respect.

Positions of Metals in the Electronegativity Series

The second factor determining the possibility of reactions having metallic character is the positions of the metals in the electronegativity series of elements. In chemistry the concept of electronegativity involves [30-36] all those properties of the elements which are connected with the behavior of the outer electrons in the atoms. As a rule, an element is electropositive if it has a relatively strong tendency to give up an electron or electronegative if it has relatively high electron affinity. The difference in electronegativity between atoms A and B in chemical bond formation may be regarded as the degree of transition of electrons from atom A to B or vice versa. A measure for the quantitative estimation of electronegativity is the energy required to remove an outer electron from one atom A and attach it to the other atom B. The concept of electronegativity and its mathematical expression have been worked out by many authors — Pauling [30], Gordy [31], Haissinsky [32], and others — who proposed equations expressing the value of electronegativity [33]. We cannot give a detailed exposition of the concept of electronegativity in chemistry or discussion of it here; it is set forth and discussed quite fully in [30-35].

Electronegativity values of the elements, calculated by the three authors cited above, are shown in Table 1. For many elements these values agree, but there are discrepancies for several. It may be concluded from these data that metallic elements are most electropositive and nonmetals most electronegative.

In connection with the present book, we note, without considering disputed questions regarding the concept of electronegativity [33-35], that this concept is important in general chemistry and, particularly, metal chemistry. It enables one to determine the character of interaction of metals and to understand the nature of chemical bonding in various types of metallides. We shall use it in this book.

In a qualitative consideration of this question, the position of the metals with respect to electronegativity is shown by an arbitrary series in which each preceding metal is regarded as electropositive relative to the following ones, and each following metal is electronegative relative to the preceding. This arbitrary electronegativity series is given below for metals and nonmetals in accordance with the nomenclature of inorganic compounds, devised by the Commission for Nomenclature of Chemical Compounds of the Division of Chemical Sciences, Academy of Sciences, USSR [36].

Electronegativity Series of Elements of the Periodic System According to [36]

Electropositive metals: Fr; Cs; Rb; K; Na; Ba; Sr; Ca; Li; Mg.

Transition metals: Ac; La; — lanthanides — Y; Sc; Th; Hf; Zr; Ti; Pa; Ta; Nb; V; — actinides — W; Mo; Cr; Re; Os; Ir; Pt Tc; Ru; Rh; Pd; Mn; Fe; Co; Ni.

Electronegative metals: Au; Ag; Cu; Hg; Cd; Zn; Tl; Pb; In; Ga; Al; Be; Bi; Sn; Sb; Ge.

TABLE 1. Electronegativity of Elements in the Periodic System According to Pauling [30], Gordy [31], and Haissinsky [32]

Group No.	Element	Electronegativity according to			Group No.	Element	Electronegativity according to		
		[30]	[31]	[32]			[30]	[31]	[32]
IA	Li	1.0	1.0	1.0	IB	Cu	2.0	1.9	2.0
	Na	0.9	0.9	0.9		Ag	1.9	1.8	1.8
	K	0.8	0.8	0.8		Au	2.4	2.1	2.3
	Rb	0.8	0.8	0.8					
	Cs	0.7	0.7	0.7					
	Fr	0.7		~0.7					
IIA	Be	1.5	1.5	1.5	IIB	Zn	1.6	1.5	1.5
	Mg	1.2	1.2	1.2		Cd	1.7	1.5	1.5
	Ca	1.0	1.0	1.0		Hg	1.9	1.8	1.9
	Sr	1.0	1.0	1.0					
	Ba	0.9	0.9	0.85					
	Ra	0.7		0.8					
IIIA	Sc	1.3		1.3	IIIB	B	2.0	2.0	2.0
	Y	1.2		1.2		Al	1.5	1.5	1.5
	La—Lu	1.1—1	1.1	1.1		Ga	1.6	1.6	1.6
	Ac	1.1		~1.0		In	1.7	1.6	1.6
	Ce		1.1	1.05		Tl	1.8	1.7	1.5
	Pr		1.1	1.1					
IVA	Ti	1.5	1.6	1.6	IVB	C	2.5	2.5	2.5
	Zr	1.4	1.4	1.4		Si	1.8	1.8	1.8
	Hf	1.3	1.4	~1.3		Ge	1.8	1.8	1.7
	Th	1.3	1.1	1.1		Sn	1.8	1.7	1.65
						Pb	1.8	1.7	1.6
VA	V	1.6	1.6	1.35	VB	N	3.0	3.0	3.0
	Nb	1.6	1.6	~1.6		P	2.1	2.1	2.1
	Ta	1.5	1.4	~1.4		As	2.0	2.0	2.0
	Pa	1.5		~1.4		Sb	1.9	1.8	1.8
						Bi	1.9	1.8	1.8
VIA	Cr	1.6	1.6	~2.1	VIB	O	3.5		3.5
	Mo	1.8	1.6	~2.1		S	2.5	2.5	2.5
	W	1.7	1.7	2.1		Se	2.4	2.3	2.3
	U	1.7	1.3	1.3		Te	2.1	2.1	2.1
						Po	2.0		2.0
VIIA	Mn	1.5	1.7	~2.3	VIIB	H	—	—	2.1
	Tc	1.9				F	4.0		4.0
	Re	1.9	1.9			Cl	3.0		3.0
						Br	2.8		2.8
						J	2.5		2.6
						At	2.2		—
VIIIA	Fe	1.9	1.7	1.8					
	Co	1.8	1.7	1.7					
	Ni	1.8	1.7	1.7					
	Ru	2.2	2.0	2.05					
	Rh	2.2	2.1	2.1					
	Pd	2.2	2.0	2.0					
	Os	2.2	2.1	~2.1					
	Jr	2.2	2.1	2.1					
	Pt	2.2	2.1	2.1					

Electronegative nonmetals: Si; B; Te; As; P; C; H; At; Se;S; N; I; Br; Cl; O; F.

Comparison reveals some discrepancy between the positions of elements in this series and the electronegativity values given in Table 1. As is evident from the series, the alkali and alkaline-earth metals (except beryllium) are the most electropositive, whereas the B-subgroup metals and beryllium are the most electronegative. Of the latter, Al, Be, Bi, Sn, Sb, and Ge lie ahead of the nonmetals at the end of the series; this illustrates the fact that they are the most electronegative metals. Thus, as noted above, beryllium lies alongside aluminum with regard to this important chemical property. In conformity with this, investigations of beryllium alloys show that, being a more electronegative element than magnesium, it is more like aluminum than magnesium in the formation of solid solutions and metallic compounds.

These two extreme positions of metals and nonmetals in the electronegativity series are the main factor determining the formation of a broad class of simple metal compounds. Between the electropositive and electronegative metals lie the transition metals, which have an unfilled d-electron shell. This series begins with actinium and ends with nickel, the most electronegative metal in this group.

Valence and Ionization Potential

As is well known, the main factor determining chemical reactions between elements and the compositions of their compounds is their valence. In many cases the number of valence, or outer electrons determines the atomic ratios in the chemical compounds formed in these reactions.

The number of valence electrons and their quantum state in the interacting metals plays an important part in establishing the maximum concentrations of solid solutions and the compositions of metallic compounds. As do the properties given above, the valence of elements depends on their position in the periodic system. The most electropositive metals (Group I) each have one outer electron; the number of these electrons increases with

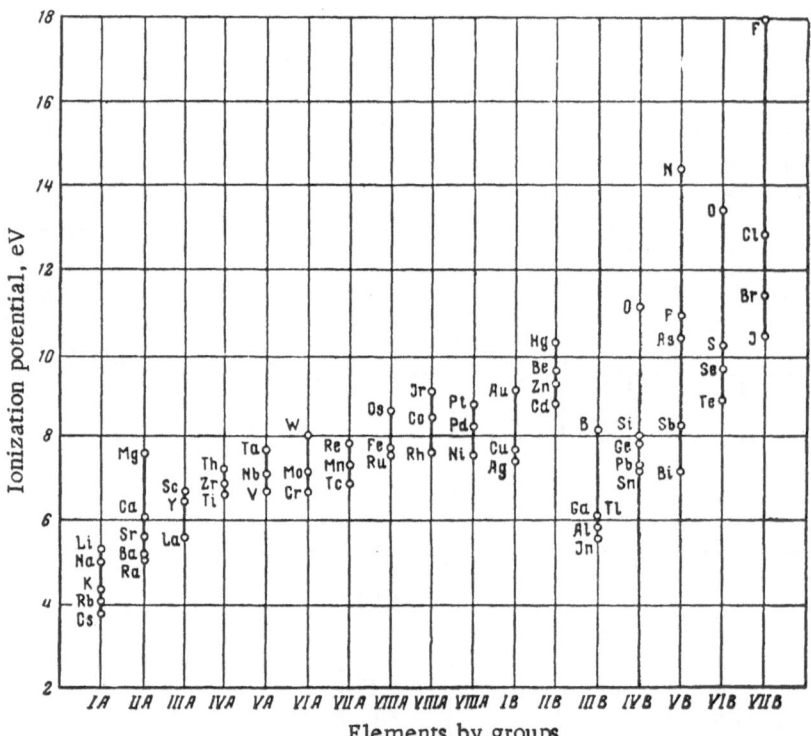

Fig. 2. Ionization potential (first) of elements by groups in the periodic system.

group number up to six for metals and seven for nonmetals. The simplest interrelations are those of isovalent metals lying in the same group; these interrelations grow more complex in heterovalent systems as the difference in the number of outer electrons in the metal and nonmetal atoms increases.

The binding strength of electrons in atoms is determined by the ionization potential, a measure of which is the energy required to remove outer (valence) electrons from the atom [17, 23, 25, 30].

The first ionization potentials of some elements are shown in Fig. 2. Here, also, they are given by groups and not by atomic number. It may be inferred from these data that the most electropositive metals of Group I have the lowest ionization potentials; the latter increase with group number, as is evident from Fig. 2. It is also characteristic that transition metals, regardless of the number of outer electrons in their atoms, have nearly identical ionization potentials. This fact plays an important part in the formation of a broad class of solid solutions based on these metals and in their relatively slight tendency to form compounds among themselves.

The highest ionization potentials correspond to the Group Zero elements, which have completely filled electron orbitals. In any one group, with some exceptions, this property decreases with increasing atomic number of the element in the given group. Metals are distinguished from nonmetals by lower ionization potentials, as is especially noticeable in the case of elements of the same group. In Group IVB, for instance, these potentials decrease markedly on passing from C to Si and less on passing from Si to the metals Ge—Sn—Pb, the number of outer electrons being the same for all five elements (see Fig. 2). The change in ionization potential is more pronounced for elements of Group VB, where nitrogen has the highest value, phosphorus and arsenic have high, nearly equal values, and antimony and bismuth have values close to those of metals. Based on these quantities, one may say that in ordinary chemical reactions electropositive metallic elements tend to give up their outer electrons easily, whereas electronegative ones — nonmetals with high ionization potentials — tend to accept electrons. These donor and acceptor properties of metals and nonmetals, manifested to a varying degree, are important in the formation of different kinds of metal compounds, having a wide variety of types of chemical bonding.

Thus the character of reactions between elements varies with increasing difference in their metallochemical properties, and the type of bonding in their reaction products shifts gradually from metallic to ionic.

CHAPTER II

METALLIDES AS INORGANIC COMPOUNDS

Classification of Compounds According to Kurnakov

Simple inorganic compounds of metals are characterized by various types of chemical bonding. Depending on the character of interaction of atoms and electron redistribution in crystals of these compounds, they have metallic, covalent, and ionic bonding or intermediate types [16-26, 30-35].

Many compounds of metals with one another or with nonmetals have metallic bonding and electronic conductivity; they are purely metallic compounds and have all the properties of metals. Numerous compounds of metals with semimetals or nonmetals have covalent bonding; many of them are semiconductor compounds.

Finally, some metals form ionic compounds with very electronegative nonmetals. As many investigations have shown, the formation of this or that type of metal compound is determined by the relative position of the elements in the periodic system, i.e., the similarity or difference in electronic structure of the reacting atoms or, as we call it, the similarity or difference in metallochemical properties of the elements.

Despite the substantial differences among inorganic metal compounds with different types of chemical bonding, they have much in common in the genetic respect, and gradual transitions from one type of compound to another are possible. These particular mutual transitions and the formation of intermediate types of bonding are characteristic of compounds with metallic and covalent bonding or covalent and ionic bonding. These questions are discussed in detail in Pauling's monograph [30].

In a number of metal—metal or metal—nonmetal equilibrium systems, as will be shown below, the metal compounds formed there have metallic, covalent, or even ionic bonding, depending on the character of interaction.

Thus various types of inorganic compounds of metals are genetically interrelated and may be named by a general term — metallic compounds, or simply metallides.

This term, first proposed by Kurnakov (see above), encompasses a larger number of simple metal compounds than the usual name "intermetallic compounds" [16-23]. The very term "intermetallic compound" literally means "compound between metals," and its meaning excludes compounds of metals and nonmetals.

The name "metallic compounds" or "metallides," proposed above, conforms to the recommendations of the Commission on the Nomenclature of Inorganic Compounds of the Division of Chemical Sciences, Academy of Sciences, USSR [36]. Among these recommendations for the group of simple compounds of metals and nonmetals, such names are proposed as borides, carbides, silicides, etc., among which compounds both with metallic and covalent bonding are encountered. In analogy with this, one may use the names aluminides, arsenides, beryllides, bismuthides, plumbides, selenides, stannides, sulfides, tellurides, and other metal compounds, which also have metallic or covalent bonding in many cases. To this group one may refer certain compounds of metals with oxygen and hydrogen — oxides and hydrides, which also have metallic or covalent bonding (see below).

Thus the word "metallides" may be used in metal chemistry as a term combining various classes of simple inorganic compounds of metals in studying reactions of metals both with other metals and with nonmetals.

Besides the papers cited above [24-29], many other studies [37-42] have dealt with the formation of metallides, questions of their classification and structure, and also a review of investigations in this field.

9

According to Kurnakov [8, 14, 15], proceeding from a general chemical viewpoint, metallides may be divided into two principal types: 1) compounds of variable composition (berthollides) and 2) compounds of constant composition daltonides.

These two types of metal compounds are now generally recognized, and these names are widely used in inorganic chemistry, metallurgy, and metal physics. They encompass not only metallic compounds formed on crystallization from melts, but also compounds resulting from transformations of solid solutions (see below). The berthollides and daltonides include a large group of metal compounds containing nonmetals with small atomic radii, such as carbon, nitrogen, oxygen, and hydrogen. Many of the compounds known as carbides, nitrides, and, in part, oxides and hydrides are interstitital phases and have typically metallic properties.

The classification of refractory compounds, among which metallides of transition metals predominate, is discussed in [42], the wide variety of such compounds being taken into account. Based on chemical criteria of formation of simple inorganic compounds, the author proposes that they be divided into three classes:

1) mutual compounds of metals (metallides proper, or intermetallides);
2) metal-like compounds (carbides, nitrides, phosphides, etc.);
3) mutual compounds of nonmetals (nonmetallic compounds of boron and carbon, carbon and silicon, boron and nitrogen, etc.).

The proposed classification is based also on the rules of change in character of chemical bonding with respect to position of the elements in the periodic system. The first class comprises compounds of metals with metals — metallides in the strict sense; the following compounds may be among them: aluminides, antimonides, beryllides, stannides, and many others. In the second class, too, there are many compounds having metallic bonding. As a rule, compounds of the first and second classes have relatively simple crystal structures with cubic, hexagonal, or tetragonal lattices and close packing of atoms.

All these compounds are the object of metallochemical investigation. Some metal compounds of the second class, containing more nonmetal than metal atoms, do not have purely metallic bonding. In them a mixed type is observed in which covalent or even ionic bonding predominates. Many of them have complex crystal structures and exhibit semiconductor properties, high electrical resistivity, and other special physical properties.

The third class of compounds includes refractory mutual compounds of nonmetals such as boron and carbon, carbon and silicon, silicon and boron, etc. They are characterized by covalent bonding and the absence of metallic elements or, therefore, metallic bonding. All such compounds occur in equilibrium systems as independent components and, except for berthollides, are distinguished on property — composition diagrams by singular points. The feature of singularity in properties is characteristic of compounds having constant composition, whether formed by crystallization from melts or by transformation in solid solutions.

The existence of metallic compounds in equilibrium systems is expressed by the three sets of property — composition and equilibrium diagrams shown in Fig. 3abc. Many such equilibrium diagrams of metallic systems are given in [43].

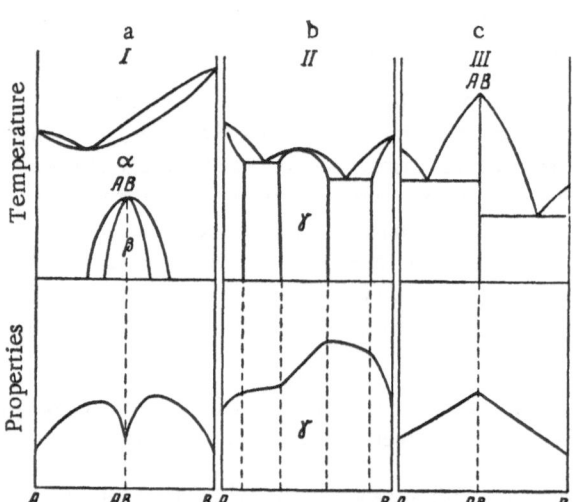

Fig. 3. Diagrams of formation of metallic compounds in binary systems: a) Kurnakov compounds; b) berthollide-type compounds; c) daltonide-type compounds.

The first of them shows the formation of compounds from solid solutions, whereas the second and third show the formation of bertholides and daltonides on crystallization. These diagrams, which indicate the formation of such compounds, will be considered in general below.

Kurnakov Compounds

The first equilibrium diagram (Fig. 3a) corresponds to the formation of metallides from primary solid solutions. As is well known, many solid solutions on slow cooling or prolonged annealing form compounds having ordered structure. They were first discovered (1914) by Kurnakov in the system $Au-Cu$[8, 37]. In this study it was noted that the copper-gold system is an example of the formation of definite compounds—gold cuprides Cu_3Au and $CuAu$ — on decomposition of continuous solid solutions. The latter play the same part with respect to these compounds as liquid solutions do in preparing most individual substances. Later such compounds with ordered structure were found in many solid solutions [26, 31], and they were called Kurnakov compounds in honor of their discoverer [26].

The hypotheses advanced earlier, that compounds are formed from solid solutions in accordance with the phase rule and that they may be regarded as independent phases with interfaces between the ordered and disordered phases, were confirmed in several recent investigations [29, 44, 45]. This justifies the plotting of equilibrium diagrams (Fig. 3a) showing such compounds.

Systems with continuous solid solutions, in which Kurnakov compounds are not formed, may not be regarded as exceptional (26, 31). In several reports to the symposium mentioned above [29], these metallic formations also were attributed to independent phases—compounds. In equilibrium diagrams they have regions of existence, and their compositions are represented by singular points. Many individual phases formed in continuous solid solutions based on Group VIII metals are compounds of a similar kind. The well-known compounds called σ-phases are formed in iron-based systems [43].

Metallides of this class are formed both in continuous solid solutions and in systems having limited regions of solid solutions. Examples of such compounds are: VNi_3, VCo_3, $MnAu_3$, $MgAg_3$, Au_3Zn, Cu_3Zn, Fe_3Al, Ni_5Sb, Ti_6Al, Ti_3Al, Ti_6O, Ti_3O, and many others. The compositions and structures of these compounds, formed from limited solid solutions of the corresponding systems, are described in [43].

To illustrate cases of compound formation from limited solid solutions, we give here new data on the equilibrium diagrams of the titanium—aluminum and titanium—oxygen systems, which we obtained in recent investigations. According to earlier literature data [43], the phase diagram of the system $Ti-Al$ contained a broad region of α- and β-solid solutions, and aluminum raised the polymorphic $\alpha \rightleftharpoons \beta$-transition temperature of titanium (Fig. 4). Subsequent detailed investigations by various authors [43] showed that in this system phase transition occurs in the region of α-solid solutions of titanium, and various forms of the phase diagram were proposed. A detailed analysis of literature data on the titanium—aluminum system is given in [46, 47].

In 1961 a paper was published [46] on the results of investigating the titanium—aluminum system by Hall-effect measurements. It was found experimentally that the Hall-effect—composition diagram shows breaks in the curves and property maxima in the region of α-solid solutions and the compound TiAl.

As the bottom of the diagram in Fig. 4 shows, the first break in the curve occurs at the composition 14.3 mole %(9 wt. %) Al; the most pronounced maximum on the curve occurs at 25 mole %(16 wt. %) Al, and the third break corresponds to 50 mole %(36.02 wt. %) Al. These compositions, at which there are breaks or maxima, correspond to the compounds Ti_6Al, Ti_3Al, and $TiAl$. Of these compounds, TiAl was known earlier, whereas Ti_6Al and Ti_3Al were found by Hall-effect measurements and some other methods. Detailed investigations of alloys of this system by structure analysis and other methods are now in progress.

One example of the discovery of new compounds formed on decomposition of solid solutions is an investigation of the equilibrium diagram for the system $Ti-O$ and proof of the existence of the two new compounds Ti_6O and Ti_3O in this system [48].

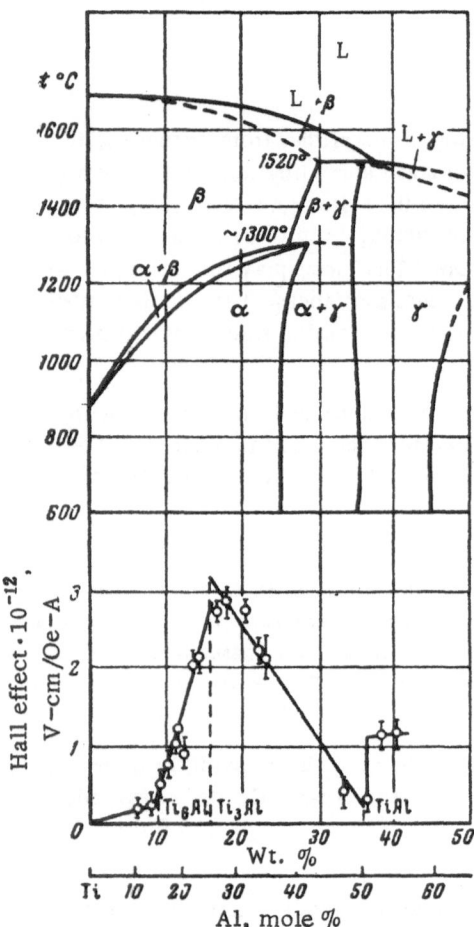

Fig. 4. Equilibrium diagram of system Ti—Al according to literature data, and Hall-effect—composition curve showing new compounds Ti$_6$Al and Ti$_3$Al.

Until now the character of interaction of titanium and oxygen and the equilibrium diagram of this system [43] were based on the supposed presence of a broad region of limited α-solid solutions of titanium and oxygen and the existence of the compounds TiO, Ti$_2$O$_3$, and TiO$_2$ in the system (Fig. 5).

The variation of physicochemical, mechanical, and other properties of alloys of the titanium—oxygen system also was based hitherto on a supposed broad region of these interstitial α-solid solutions, extending to 32 mole %O.

Statements regarding the possible formation of an ordered structure in the region of α-solid solutions of titanium and oxygen have appeared in some papers in recent years.

Based on analysis of data on reactions of Group IV metals (Ti, Zr, Hf, Th, C, Si, Ge, Sn, and Pb) with oxygen, the author in 1956 proposed the possible formation of the compound Ti$_3$O in this system. This is confirmed especially by the presence of a flat maximum in the fusibility diagram of the titanium—oxygen system, corresponding to ~25 mole %O (see Fig. 5).

Owing to the theoretical and practical importance of the interaction of titanium and oxygen, as well as that of finding out the chemical nature of alloys of the titanium—oxygen system, this system was investigated in detail at concentrations from 0 to 35 mole %O [48]. Oxygen was introduced in the form of a titanium—oxygen alloy containing 15.8 wt. % oxygen. This alloy was prepared by melting rods, pressed from titanium and titanium dioxide containing 99.93% TiO$_2$, in an arc furnace.

The cast alloys were homogenized in vacuo at 800° for 1000 hr. After this the alloys were further annealed at 400, 600, 800, 850, 1000, and 1400° for 600, 400, 1000, 100, and 2 hr, respectively, and then quenched in ice water.

Chemical reactions and the structure of the equilibrium diagram were studied by physicochemical analysis: microscopic and x-ray structure analysis, as well as measurement of microhardness, electrical resistivity, and thermo-emf.

Experimental investigation of alloys of the titanium—oxygen system showed that the two compounds TI$_6$O and TiO$_3$, previously unknown in the TI—O system, are formed from α-solid solutions of titanium [48]. Both compounds have metallic properties and the same crystal lattice (hexagonal); the x-ray diffraction pattern of Ti$_3$O contains superstructure lines, which indicate that this compound has an ordered structure. TiO, Ti$_2$O$_3$, and TiO$_2$, known earlier, are semiconductor- and valence-type compounds. In the case of titanium—oxygen compounds it is quite obvious that the type of chemical bonding changes gradually from metallic to ionic with increasing number of oxygen atoms in the titanium oxides (Ti$_6$O → Ti$_3$O → TiO → Ti$_2$O$_3$ → TiO$_2$). Ti$_6$O and Ti$_3$O are Kurnakov compounds; as yet they are the only examples of oxide formation from limited solid solutions of metals. The oxygen atoms in these compounds apparently are ordered as in interstitial compounds. On the basis of these data, Hansen's equilibrium diagram of the system Ti—O [43] (Fig. 5), having a broad region of α-solid solutions of titanium, is superseded in [48] by an equilibrium diagram (Fig. 6) in which two new compounds, Ti$_6$O and Ti$_3$O, and their regions of existence are shown. Both these compounds, formed from α-solid solutions, are characterized by singular points on the property—composition diagrams [48].

Judging from the character of the property—composition diagrams for 800° and from data of microscopic analysis, one may conclude that Ti_6O is stable up to 820-830°.

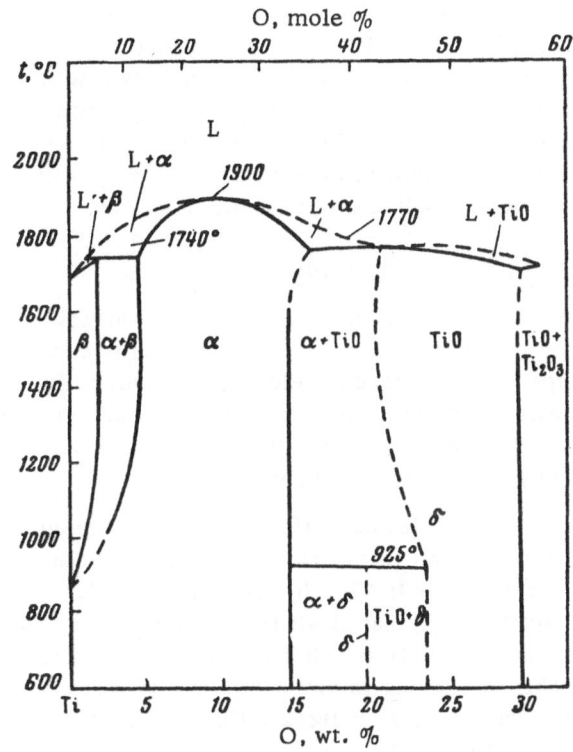

The compound with an ordered structure (Ti_3O) apparently is stable up to 1400° or more, since the singular character of the property—composition diagram at this temperature remains unchanged. According to recent data of the authors, it is stable up to the melting point (1940°).

Proceeding from the analogy of titanium with zirconium and hafnium, one may presume that similar compounds can be formed in the system Zr—O and Hf—O, which are analogous to the system Ti—O. Further investigations of these systems may reveal that new compounds of the following possible compositions are formed in them from α-solid solutions: Zr_6O and Zr_3O in the system Zr—O; Hf_6O and Hf_3O in the system Hf—O. They should be isomorphous with the compounds Ti_6O and Ti_3O.

Fig. 5. Equilibrium diagram of system according to literature data.

The discovery of such oxygen compounds of titanium and its analogs is very important for understanding the structure and physicochemical nature of alloys based on titanium, zirconium, and hafnium, containing a high percentage of oxygen, and for finding out the effect of oxygen on the mechanical and other properties of titanium, zirconium, and hafnium.

Detailed investigations of many metallic systems containing continuous and limited solid solutions should result in the discovery of numerous new Kurnakov compounds, which have not yet been found in homogeneous solid solutions.

Berthollide-Type Compounds

Metallides of variable composition (berthollides), formed by crystallization, are characterized by the presence of regions of homogeneity and the absence of well-defined singular points in the property—composition diagrams, corresponding to the composition ordinates of compounds (see Fig. 3b). Among metallic compounds in general, berthollides lie between Kurnakov compounds and daltonides with respect to stability. There are cases where berthollides stable at high temperatures go over to Kurnakov compounds with ordered structure on cooling. A striking example of this type in the system Cu—Zn is the transition of the high-temperature β-phase (berthollide) to the β'-phase, i.e., the Kurnakov compound CuZn, on cooling.

The compositions and crystal structures of such compounds are often determined by the ratio of the number of outer electrons to that of atoms in the structure. They are called electron compounds. The atomic ratios in such compounds are variable; they are determined by the ratio e/a of the number of outer electrons to the number of reacting atoms [19, 20]. These phases are divided according to these ratios into a β-phase with the e/a ratio 3/2 (e.g., CuZn), a γ-phase with the ratio 21/13 (e.g., Cu_5Zn_8), and an ε-phase with the ratio 7/4 (e.g., $AgCd_3$). The quantum theory of such compounds has been developed by many authors [19, 20, 49]. Jones and Mott, Konobeevskii, Hume-Rothery, et al. have made especially large contributions in this respect. This type of compound is formed mainly by the interaction of subgroup B metals lying close together in the periodic system and differing slightly in electronegativity.

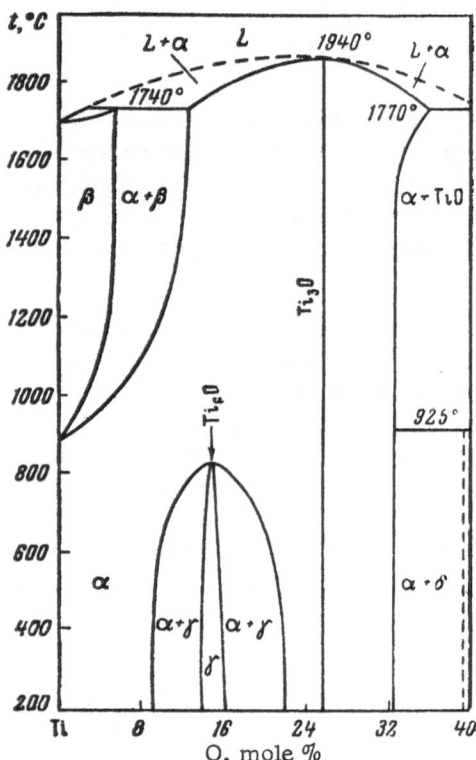

Fig. 6. Equilibrium diagram of system Ti—O, showing the new compounds Ti$_6$O and Ti$_3$O.

One may point out specific examples of the formation of such phases in systems based on Group I metals (Cu, Ag, Au) and containing Group II ones (Zn, Cd, Hg), based on Group III metals (Al, In, Ga) and containing Group V ones (Bi, Sb), etc. In some cases these metals form compounds by reacting with transition metals. Phase diagrams of such systems are described in a monograph [43]. In the region of these compounds of variable composition the properties vary with composition along a smooth curve, as shown in Fig. 3b.

Daltonide-Type Compounds

Among metals the tendency to form the second type of metallic compounds — daltonides — increases with difference in chemical properties of the reacting elements in conformity with the increasing differences in electronegativity and ionization potential. In property—composition diagrams they are characterized by singular points, as shown in Fig. 3c.

This is especially characteristic of reactions of the most electropositive metals of Groups I and II with the most electronegative metals of Groups II-VI, subgroup B. Kurnakov's remarkable hypotheses [8] that alkali and alkaline-earth metals, being the chief bearers of basic or metallic properties, are most capable of forming various compounds with heavy (most electronegative) metals — Hg, Zn, Cd, Pb, Sn, Bi — are justified in this respect.

Characteristic examples of the formation of compounds of variable composition (berthollide-type) and constant composition (daltonide-type) are the equilibrium diagrams of two typical metal systems, shown in Fig. 7ab.

One of them (Fig. 7a) is the mercury—cadmium system, composed of analogous metals located in Group IIB and having nearly identical metallochemical properties. As is evident from the figure, limited solid solutions based on cadmium and mercury are formed in this system; moreover, there is a broad region of a variable-composition phase ω, which corresponds to a berthollide. The other is the potassium—mercury system (Fig. 7b), composed of electropositive potassium and electronegative mercury. The difference in metallochemical properties between these metals is manifested, firstly, in the absence of solid solutions based on the components, secondly, in the formation of a large number of compounds, and thirdly, in the very high proportion of the electronegative metal in the compositions of the compounds.

The five compounds formed in the mercury—potassium system (KHg$_{11}$, KHg$_8$, KHg$_4$, KHg$_3$, KHg$_2$, and KHg) have higher melting points than their components; they have metallic bonding and are typical daltonide-type metallides.

In many other cases of metallic systems one can demonstrate rules of formation of berthollides and daltonides, depending on the relative positions of the reacting elements in the periodic system.

Strength and Types of Chemical Bonding in Metallides

The formation of metallic compounds and their relative stability depend on the difference in metallochemical properties, atomic radii, electronegativity, and ionization potentials of the reacting elements (see above). The stability of metallides is determined by the strength of their chemical bonding, which can be estimated, for instance, from their heats of formation. These questions are considered in a special monograph on the thermochemistry of inorganic compounds [50], a special book[51]on heats of formation and types of chemical bonding in inorganic crystals, and N. V. Ageev's paper "The phase diagram as an expression of interatomic interaction" [52].

Analysis of the literature shows that compounds of metals with analogous metals, in which the different kinds of atoms present have nearly identical metallochemical properties, are the least stable. From this viewpoint the least stable compounds are Kurnakov ones, having ordered structure, which are formed from solid solutions of analogous metals. For instance, the heat of formation of the above-mentioned compound AuCu, formed from solid solutions of the system Cu−Au, is 0.37 kcal/mole, whereas that of FeCr, formed from solid solutions of the system Fe−Cr, is 0.78 kcal/mole.

Owing to the considerable theoretical interest of this question, a special investigation was conducted [45] and the rules of change established for the heats of formation of nickel compounds of the type $MeNi_3$ with a series of metals in which the electronegativity difference relative to nickel gradually increases.

The heats and temperatures of formation (or dissociation) of nickel compounds with the metals Fe, Mn, Cr, V, and Ti, which lie in the same period of the periodic system but in different groups (VIII, VII, VI, V, and IV) are shown in Fig. 8. In conformity with the increasing difference in properties between the more electronegative nickel and the more electropositive metals Fe, Mn, Cr, V, and Ti (see Table 1), the stability of nickel compounds of a given composition ($MeNi_3$) with these metals increases in that order.

As is evident from the curves of Fig. 8, the heat and temperature of dissociation of these compounds (except $CrNi_3$) increase in the order shown in Table 2.

The considerable deviation from the curves in the instance of $CrNi_3$ requires further study. In this case it is characteristic that the first four compounds ($FeNi_3$, $MnNi_3$, $CrNi_3$, and VNi_3) are formed from solid solutions and are Kurnakov compounds, whereas TiNi, the most stable of all these compounds, is a daltonide and is formed by crystallization from the melt. It will be noted that the difference in conditions of formation (from solid solutions or melts) does not affect the order of change of their properties. This proves the generality of the rule, which we established, that the stability and the order of change in strength of chemical bonding in metallic compounds depend on the position of the reacting elements in the periodic system. This type of dependence of the relative stability of metallic compounds on the difference in metallochemical properties of the elements can be demonstrated not only in the case of Kurnakov compounds, but also in that of many berthollides and daltonides.

In daltonides formed by elements having markedly different metallochemical properties, the type of chemical bonding is changed. Among these compounds it is easy to follow the gradual transition from metallic through covalent to ionic bonding and to show the predominance of this or that type of chemical bonding in a series of many such compounds. This gradual transition takes place as the electronegativity of one component increases. Many

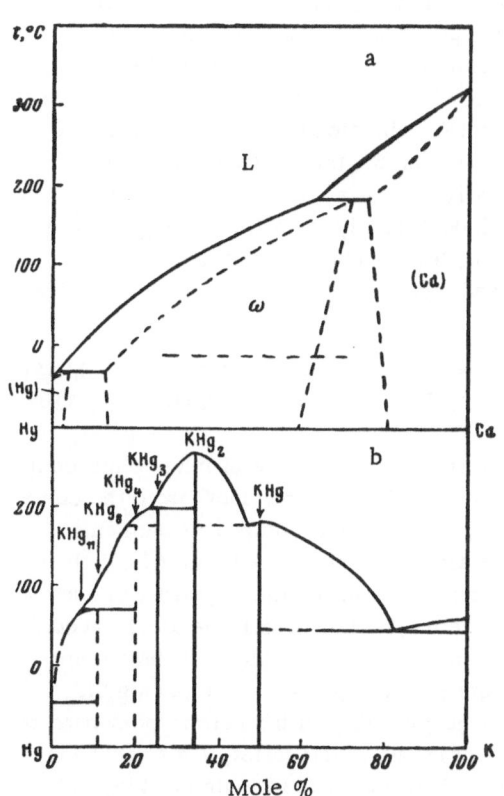

Fig. 7. Binary metal systems; a) Kg−Cd with berthollide phase ω; b) Hg−K with daltonide-type compounds.

TABLE 2

Property	Compound				
	FeNi$_3$	MnNi$_3$	CrNi$_3$	VNi$_3$	TiNi$_3$
Heat of formation ΔH, kcal/mole	1.97	2.40	0.41	3.60	8.4
Dissociation temperature, °C	570	540	585	1070	1380

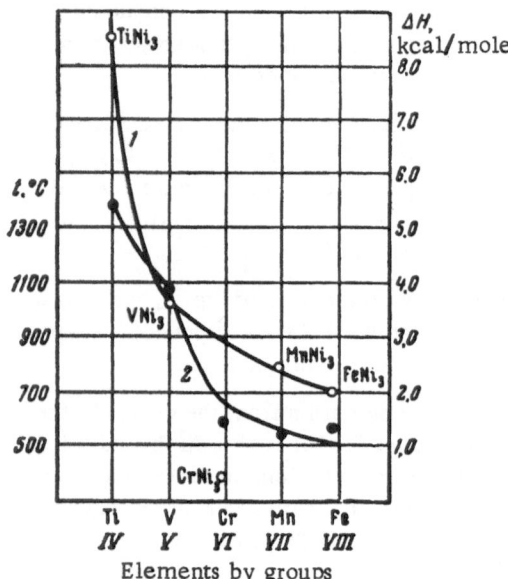

Fig. 8. Heats temperatures of formation of Ni compounds with Fe, Mn, Cr, V, and Ti: 1) heat of dissociation; 2) dissociation temperature.

compounds of electropositive magnesium and other, electronegative elements, exhibit transition from metallic to covalent and ionic bonding with increasing differences in metallochemical properties of the elements reacting with magnesium. This may be observed in a number of magnesium compounds with elements of Group IVB: Pb, Sn, Ge, and Si. The compositions of these compounds correspond to the general formula Mg_2X. As is evident from Fig. 9, the equilibrium diagrams of the systems Mg—Pb, Mg—Sn, Mg—Ge, and Mg—Si show the formation of compounds having identical atomic ratios: Mg_2Pb, Mg_2Sn, Mg_2Ge, and Mg_2Si. They all conform to the laws of valence and have open melting-point maxima. These daltonides differ from other compounds by the fact that they do not form solid solutions with the components. As the electronegativity of the elements increases from the metal Pb to the nonmetal Si, the melting point of the compounds gradually increases, and a transition takes place from compounds having purely metallic bonding, such as Mg_2Pb, to covalent, semiconductor compounds such as Mg_2Sn, Mg_2Ge, and Mg_2Si. Their melting points are given below:

Compound	Mg_2Pb	Mg_2Sn	Mg_2Ge	Mg_2Si
Melting point	550	778	1070	1102

Such a transition and gradual change in type of bonding occur in the following series of magnesium compounds with third-period elements: $Mg_3Al_2 \rightarrow Mg_2Si \rightarrow Mg_3P_2 \rightarrow MgS \rightarrow$ and $MgCl_2$. As is evident from the compositions of these compounds, they also conform to the rules of valence, whereas their component elements (Al, Si, P, S, and Cl) increase in electronegativity and ionization potential (see Table 1 and Fig. 2). Of these elements, only Al is a metal, whereas the rest are nonmetals. Although Al lies in the third period, as does Mg, its electronegativity is considerably different from that of electropositive magnesium (see Table 1). In accordance with this, aluminum forms a number of compounds with magnesium, including Mg_3Al_2. All these compounds are typically metallic; the following three compounds with more electronegative elements (Mg_2Si, Mg_3P_2, and MgS) have covalent bonding and are semiconductor compounds, possibly with a certain proportion of ionic bonding, especially in magnesium phosphide and sulfide. $MgCl_2$, the last in this series, is a typical ionically bonded compound. This example confirms the gradual transition from metallic to ionic bonding with increasing difference in electronegativity and ionization potential of the metals and nonmetals reacting with magnesium.

The strength of chemical bonding in magnesium compounds increases with the difference in metallochemical properties between magnesium and the series of elements listed above, either by groups or by periods.

TABLE 3

Compounds with Group IVB elements	Q, kcal/mole	Compounds with Group VIB elements	Q, kcal/mole	Compounds with third-period elements	Q, kcal/mole
Mg_2Pb	4.2	MgTe	25.0	Mg_2Si	6.3
Mg_2Sn	6.1	MgSe	30.0	Mg_3P_2	25.6
Mg_2Ge	6.1	MgS	42.2	MgS	42.2
Mg_2Si	6.3	MgO	73	$MgCl_2$	51.1

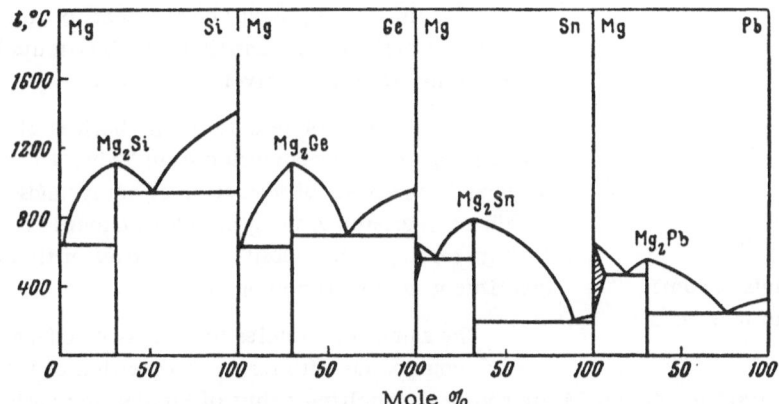

Fig. 9. Phase diagrams of systems Mg—Pb, Mg—Sn, Mg—Ge, and Mg—Si, showing compositions of compounds Mg_2Pb, Mg_2Sn, Mg_2Ge, and Mg_2Si.

This gives rise to increased heat of formation and melting points for a given stoichiometric composition. A number of rules connected with this are pointed out in [40, 50-52]. These questions are considered in the case of magnesium compounds with elements of Groups IVB and VIB, as well as those of the third period, in the cited paper [52].

Data on the change in heats of formation Q for magnesium compounds with elements having successively increasing electronegativity difference (relative to magnesium) are given in Table 3.

Acquaintance with the heats of formation of such compounds reveals that intermetallides, formed by the reaction of a metal with a metal, have much lower heats of formation than compounds of metals with non-metals, having characteristic covalent or ionic bonding. In all three columns the heat of formation increases from top to bottom in accordance with the increase in electronegativity for elements of Group IVB from Pb to Si, Group VIB from Te to O, and the third period from Si to Cl.

Consideration of the order of this change by groups and periods shows that in Groups IVB and VIB the heat of formation of magnesium compounds increases with decreasing atomic number, whereas in the third period (from Si to Cl), on the contrary, it increases with increasing atomic number. In either case this corresponds to increasing difference in electronegativity between magnesium and the reacting elements.

The stability of metallides also varies with their melting point. As a rule, the maximum strength of chemical bonding corresponds to the maximum melting point. This is illustrated by the phase diagrams of the systems Mg—Pb, Mg—Sn, Mg—Ge, and Mg—Si (see Fig. 10). In accordance with the above equilibrium diagrams and heats of formation of the compounds Mg_2Pb, Mg_2Sn, Mg_2Ge, and Mg_2Si, their melting points increase regularly. The melting points of lithium compounds with lead and tin also are plotted from literature data [43] in Fig. 10 in analogy with Mg compounds. These points also lie on a straight line and show that the melting point of the compound increases from Li_4Pb to Li_4Sn.

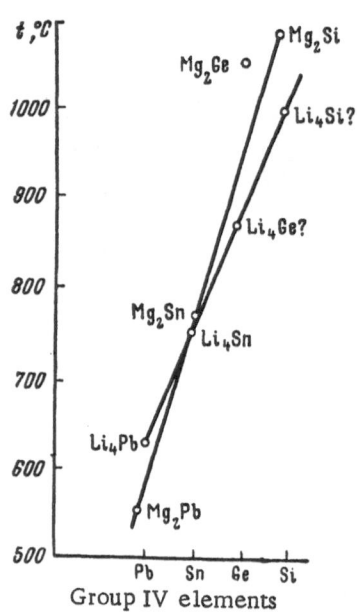

Fig. 10. Melting points of some metallic compounds of magnesium and lithium.

By extrapolating the straight line to the hypothetical compounds Li_4Ge and Li_4Si (there are no literature data), one can determine the melting points of these compounds. Presumably they will be $\approx 870°$ for Li_4Ge and $\approx 980°$ for Li_4Si.

The rules of change in the heats of formation and melting points of metal compounds with respect to position of the reacting elements in the periodic system apply to many metallides. Such rules were established by studying successive changes in type of chemical bonding, heat of formation, and melting point in the case of borides [53], carbides [54], silicides [55], phosphides [56], and other compounds of metals and nonmetals.

Literature data [53-55] on the heats of formation $\Delta H°_{298}$ of some borides and carbides of transition metals (Groups IV and V), having the same atomic ratios, are given in Table 4.

Although complete data on the borides of these metals are lacking, one can draw a conclusion from Table 4 regarding successive changes in the heats of formation of compounds of transition metals within each group. Among metals of Groups IV and V they increase with increasing electropositive properties of these metals for a given constituent nonmetal—boron or carbon.

The same rule applies to successive change in the melting points of such compounds with respect to position of the metals in separate groups of the periodic system. In Fig. 11 are shown the melting points of borides and carbides of metals in Groups IV, V, and VI, from which it is evident that for these compounds the melting point increases linearly with the atomic number of the metal. This corresponds to increase in the electropositivity of metals and the difference in electronegativity between metals, on the one hand, and boron and carbon, on the other.

Compounds of phosphorus and metals also are very interesting in this respect. In [56], which deals with phosphides, the authors consider many compounds of phosphorus and transition metals. They note that phosphorus compounds with transition metals do not constitute interstitial phases (except lanthanide phosphides), on the grounds that phosphorus has a larger atomic radius and lower electronegativity than its analog, nitrogen.

Phosphides of metals should have a far smaller proportion of ionic bonding than nitrides and less metallic character than arsenides, antimonides, and bismuthides. Moreover, phosphorus has a more marked tendency to form metallic compounds than sulfur. This is in accord with the relative positions of these elements in the electronegativity series (see Table 1).

As the number of phosphorus atoms in transition metal phosphides decreases, the role of the Me—P and P—P bonds diminishes and that of the Me—Me bond grows, so that, as in the case of carbides and nitrides, the metallic properties of the phosphides are enhanced.

In conformity with the larger proportion of metallic bonding in phosphides as compared with nitrides, the heats of formation of the former should be less than those of the latter. This is due to the lower ionization potential and electronegativity of phosphorus compared with nitrogen. The heat of formation Q of phosphides increases with the number of phosphorus atoms in them, as is evident from Table 5 in the case of iron, cobalt, and nickel phosphides [56].

TABLE 4

Composition	$-\Delta H°_{298}$, kcal/mole	Composition	$-\Delta H°_{298}$, kcal/mole
Group IV			
TiB_2	70.2	TiC	43.8
ZrB_2	78.0	ZrC	44.4
HfB_2	—	HfC	88.0
Group V			
VB_2	—	VC	28
Nb_2	36.0	NbC	33.7
TiB_2	<52	TaC	38.5

TABLE 5

Compound	Q, kcal/mole	Compound	Q, kcal/mole	Compound	Q, kcal/mole
Fe_3P	13.1	Co_2P	23.45	Ni_5P_2	17.4
Fe_2P	19,25	CoP	34.0	Ni_3P	19,2
FeP	29,0	CoP_3	64.0	NiP_2	38,0
FeP_2	42,0	—	—	NiP_3	45.2

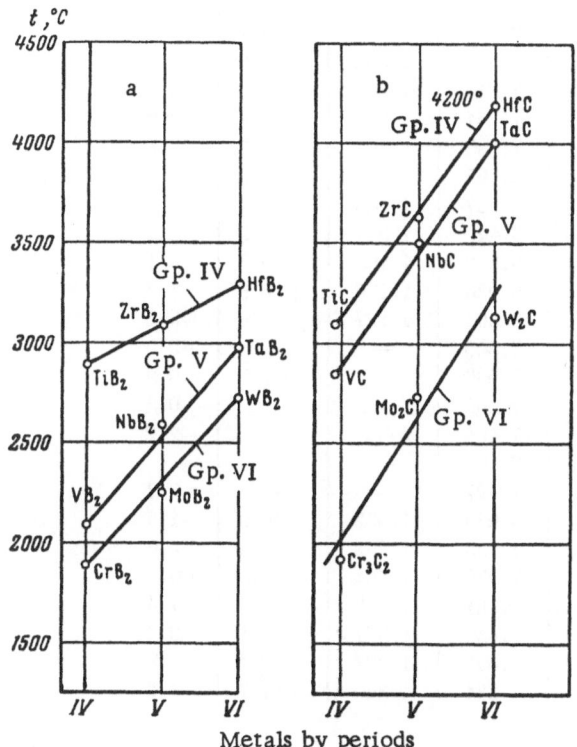

Fig. 11. Melting points of borides (a) and carbides (b) of metals in Groups IV, V, and VI.

These data attest that the number of P—P and P—Me bonds in the compounds increases with the number of phosphorus atoms, and covalent and ionic bonding begin to predominate in them, thus increasing the heat of formation of such compounds.

The above rules governing the heats of formation of transition metal phosphides apply also to transition metal compounds with other nonmetals. A certain order in the change in character of chemical bonding in them with respect to the electronic structure of of rare-earth metals and sulfur was established in a paper on the chemistry of rare-earth sulfides[57]. In this case it was noted that the bonding in such sulfides could be more nearly metallic than in the nitrides of the corresponding metals. This is explained by the fact that sulfur has a lower ionization potential than nitrogen, as do phosphorus and carbon (see Fig. 2). Hence one may expect the electrons of the metal and nonmetal to be redistributed, or collectivized, to a larger degree in the sulfides of these metals, for a given atomic ratio, than in the nitrides. Even in the sulfides, however, metallic bonding is accompanied by ionic to a certain extent. In this case it must be stated that in sulfides, analogously to phosphides, the proportion of ionic bonding increases with the number of sulfur atoms in the compounds [57].

Structure and Compositions of Metallides

Besides the above classification of metallic compounds according to two general chemical features (berthollides and daltonides), proposed by Kurnakov, there are other classification systems for such compounds. Some of them are based on type of crystal structure [39-41, 58-63]; others, on electronic structure and the nature of chemical bonding [25, 26, 30, 64, 65].

It should be recognized, however, that no unified system for classifying inorganic compounds, particularly metallides, has appeared in the literature until now. All these principles for classifying them according to chemical, structural, and other criteria are independent of one another; they are contradictory in many cases and are not generally accepted. These questions require special consideration and discussion. Nevertheless, in order to consider the structure and compositions of various metallides in detail, we propose to divide them according to criteria of structure, chemical composition, and special physical properties. Below we shall consider from this viewpoint compounds with the structure of Laves phases, compounds with the structure of nickel arsenide phases, and compounds of metals with nonmetals — borides, carbides, silicides, nitrides, oxides, and to

TABLE 6   Compositions, Lattice Parameters, Atomic-Radius Ratios $r_A/r_B = q$, and Change in Interatomic Distances A and B for Laves Phases AB_2 with Cubic Structure ($MgCu_2$-Type)

Compound	Parameter, A	q	$100 \cdot \sigma_A$	$100 \cdot \sigma_B$	Compound	Parameter, A	q	$100 \cdot \sigma_A$	$100 \cdot \sigma_B$
$AgBe_2$	6.29	1.300	− 5.7	+0.1	$NpAl_2$	7.769	—	—	−3.9
$BaPd_2$	7.937	1.582	−20.9	+2.2	$PbAu_2$	7.897	1.214	− 2.1	−3.0
$BaPt_2$	7.904	1.568	−21.2	+0.9	$PrNi_2$	7.190	1.461	−14.4	+2.2
$BaRh_2$	7.836	1.618	−21.9	+3.2	$PuAl_2$	7.815	—	—	−3.4
$BiAu_2$	7.942	—	—	−2.4	$PuCo_2$	7.061	—	—	+0.2
$CaAl_2$	8.02	1.376	−11.7	−0.8	$PuFe_2$	7.176	—	—	+2.4
$CaIr_2$	7.530	1.451	−17.1	−1.7	$PuMn_2$	7.275	—	—	—
$CaPd_2$	7.650	1.431	−15.8	−1.5	$PuNi_2$	7.15	—	—	+1.6
$CaPt_2$	7.614	1.420	−16.1	−2.8	$RhBi_2$	9.590	—	−16.7	—
$CaRh_2$	7.510	1.464	−17.3	−1.1	$SrIr_2$	7.684	1.586	−22.6	+0.3
$CeAl_2$	8.04	1.270	− 4.1	−0.5	$SrPd_2$	7.810	1.564	−21.3	+0.5
$CeCo_2$	7.15	1.457	−14.6	+1.4	$SrPt_2$	7.761	1.551	−21.8	−0.9
$CeMg_2$	8.71	1.138	+ 3.9	−3.5	$SrRh_2$	7.690	1.601	−22.5	+1.3
$CeNi_2$	7.19	1.460	−14.3	+2.2	$TaCo_2$	6.765	1.145	+ 2.3	−4.3
$CePb_2$	7.714	1.311	− 8.0	−1.5	$TaCr_2$	6.947	1.145	+ 5.3	−1.4
$CoBi_2$	9.726	—	−19.9	—	$TiBe_2$	6.44	1.301	− 3.5	+2.5
$GdFe_2$	7.43	1.431	− 9.1	+6.1	$TiCo_2$	6.692	1.160	+ 0.3	−5.1
$GdMn_2$	7.73	—	− 5.5	—	$TiCr_2$	6.926	1.160	+ 3.8	−1.7
$HfCo_2$	6.908	1.260	− 4.7	−2.0	$ThMg_2$	8.553	1.127	+ 3.0	−5.2
$HfCr_2$	7.011	1.260	− 3.3	−0.5	UAl_2	7.800	—	—	−3.5
$HfMo_2$	7.562	1.154	+ 4.3	−1.7	UCo_2	6.990	—	—	−0.8
HfV_2	7.397	1.195	+ 2.0	−0.5	UFe_2	7.060	—	—	+0.8
HfW_2	7.556	1.147	+ 4.2	−2.3	UIr_2	7.494	—	—	−2.2
KBi_2	9.50	—	−11.0	—	UMn_2	7.160	—	—	—
$LaAl_2$	8.12	1.310	− 6.0	+0.5	UOs_2	7.497	—	—	−0.7
$LaMg_2$	8.77	1.174	+ 1.5	−2.8	$ZrCo_2$	6.940	1.452	−16.9	−1.5
$LaNi_2$	7.25	1.506	−16.2	+3.1	$ZrCr_2$	7.193	1.452	−13.9	+2.0
$MgCu_2$	7 02	1.251	− 4.3	−2.7	$ZrFe_2$	7.056	1.461	−15.5	+0.7
$NaAg_2$	7.91	1.286	− 7.6	−3.0	$ZrMo_2$	7.581	1.452	− 9.3	−1.4
$NbAu_2$	7.79	1.288	− 9.1	−4.3	ZrV_2	7.430	1.377	−11.1	0.0
$NbCo_2$	6.755	1.145	+ 2.5	−4,2	ZrW_2	7.600	1.323	− 9.1	−1.8
$NbCr_2$	6.971	1.145	+ 5.8	−1.1	$ZrZn_2$	7.381	1.360	−11.7	−1.9

some degree phosphides, sulfides, etc. Besides, we should mention some groups of compounds having special physical properties: superconductor and semiconductor compounds.

In the following sections, metallides are briefly described in accordance with the above criteria.

Compounds of the Type of Laves Phases

In many metallic systems composed of Metals A and B which are mutually insoluble or slightly soluble in the solid state, compounds are formed whose compositions correspond to the general formula AB_2. These compounds are called Laves phases [66 -73]. They are all formed between metals; nonmetals play no part in them. The component metals of Laves phases lie in various groups of the periodic system. As Ageev notes [25], they are not arranged in any definite order. All except seven metals (Hg, Ga, In, Tl, As, Sb, Sn) form Laves phases. These phases are formed in the crystallization of alloys by the reaction of two metals having a definite ratio of atomic radii. In this case the ratio $r_A : r_B$ should be 1.226, whereas it actually varies from 1.14 to 1.62.

Laves phases have three types of crystal structure: $MgCu_2$ (cubic), $MgNi_2$, and $MgZn_2$ (hexagonal). The lattice parameters of these compounds are much larger than those of the pure component metals. Most Laves compounds crystallize in the $MgCu_2$ type, some in the $MgZn_2$ type, and very few in the $MgNi_2$ type.

It has been found that the average coordination number of Laves phases exceeds 12, amounting to 13, 14, 15, or even 16, and thus the atoms are more closely packed in these compounds than in the pure metals. The formation of such structures may be explained by the considerable difference in atomic radius of the reacting metals A and B, which enables the smaller atoms to lie in the spaces between the larger. The rise in density of the compounds AB_2 increases with the difference in atomic radius of the metals A and B, and this increases their stability.

A characteristic of Laves phases is the fact that atoms of the same kind in the structure are in contact, and the interatomic distances in them are shortened more than in the structures of the reacting metals. In this case A atoms are more compressible than those of B, which contract very little.

Since all compounds of the Laves-phase type are formed by reactions between metals, they are typical intermetallic compounds and have metallic bonding. Covalent bonding is rarely encountered among them. As are all metallides, they are quite brittle at room temperature but become plastic near the melting points. Some Laves phases have practical value.

Over 200 compounds of the Laves-phase type have been discovered in the last 20-25 years. As experimental data accumulate, there will be even more, since the equilibrium diagrams of those metal systems whose components should form Laves phases have by no means all been studied. An interesting and complete review of papers on the conditions of formation of Laves phases is given in [69].

Table 6 gives information, taken from recent literature data [73], on compositions, lattice parameters, atomic-radius ratios, and changes in the interatomic distances $A-A$ and $B-B$ in Laves phases with the general formula AB_2, having a $MgCu_2$-type cubic structure.

Many Laves phases have polymorphic transformations. The composition and structure of the compounds given in the table show their isomorphism. The similarity and difference in metallochemical properties of the constituent atoms of these compounds are important conditions in studying reactions between the latter.

Compounds Having the Structure of Nickel Arsenide Phases

Metallic compounds crystallizing in a hexagonal NiAs type structure with the coordination number 8 have certain properties in common. In this structure the arsenic atoms (or analogous atoms of other elements, e.g., In, Sb, Te) are located at the points of a close-packed hexagonal lattice, whereas the Ni (Fe, Co, etc.) atoms lie at the points of an independent cubic lattice. Makarov [74] showed that a system of such phases can constitute a homologous series of compounds having a variable number of atoms in the unit cell. In fact, comparison of compounds in the series $NiTe_2-NiSb-Ni_3Sn_2-Ni_2Sn$ shows that the number of nickel atoms per unit cell gradually increases (from 1 to 4, respectively), the value of this ratio for the second component being constant (2 atoms per unit cell). These changes are clearly shown in Fig. 12 [74].

As Makarov notes, one cannot assign a definite type of chemical bonding to nickel arsenide phases; however, there is no doubt that ionic bonding becomes less marked and metallic bonding more so with decreasing difference in chemical properties of the elements and increasing relative nickel content.

As noted above, this kind of gradual transition from metallic to ionic bonding is characteristic of a number of other compounds of metals with nonmetals. As the concentration of metal atoms in the compounds formed increases, metallic properties begin to predominate. In the case of nickel compounds with sulfur, for instance, the sulfide Ni_3S_2 has metallic properties, whereas the sulfides NiS and NiS_2, which are richer in sulfur, do not.

Thus nickel arsenide type compounds are an example of the pronounced effect which the chemical nature of the component elements has on the structure and properties of the metallic compounds formed.

A detailed systematic account of nickel arsenide type phases from the viewpoint of the crystallochemical peculiarities of binary and ternary metallic compounds is given in a review article [75]. Here the authors combine type TB compounds (where T is a transition metal and B a subgroup B one) having the geometrically related structures B2, B20, B31, B35, and B8 (NiAs-type) into one "family."

The formation of these structures depends on the position of the T elements in the periodic system.

Borides

Many authors have investigated the compositions and structures of borides. Their work is described in review articles and monographs [39, 40, 53, 54, 77, 78].

Owing to the relatively large atomic radius of boron (r_B = 0.87 A, see Fig. 1), it was thought that borides could not be interstitial phases. However, recent investigations proved that many of the borides studied, e.g., those of metals in Groups IV, V, and VI, are interstitial phases in which boron atoms may be located within the structure of the metal in isolation, in the form of chains and nets, or even in compounds with high boron content (MeB_4, MeB_6, MeB_{12}), in the form of three-dimensional skeletons.

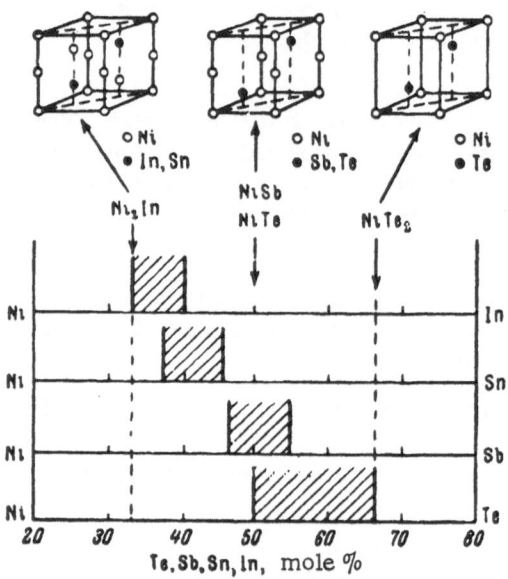

Fig. 12. Homologous group of nickel arsenide-type compounds in Ni—In (Sn, Sb, Te) systems.

The character of chemical bonding in borides depends on their composition. The bonding between boron and metal atoms is mainly metallic, which enables one to regard borides as metallic compounds. As the relative boron content increases, however, the bonding becomes less metallic and more ionic. Apparently the boron atoms themselves, which form complexes, are covalently bonded to one another.

As a quantitative estimate of the degree of metallic bonding between boron and transition metals of Groups IV, V, and VI, G. V. Samsonov [76] proposed that the ratio $1/Nn$ be used, where N is the principal quantum number of the unfilled d level and n the number of electrons in the d level. The larger $1/Nn$ is, the more probable it is that the outer electrons of boron will remain in the d level of the transition metal atoms, and the more metallic the bonding is. On the other hand, as $1/Nn$ decreases, the bonding approaches ionic.

TABLE 7

Composition of boride	Electrons in unfilled levels of metals	Ratio $1/Nn$	Electrical resistivity $\mu\Omega$-cm
TiB_2	$3d^2$	0,167	14.1
VB_2	$3d^3$	0.111	16,0
CrB_2	$3d^5$	0.067	25,0
ZrB_2	$4d^2$	0.125	10.4
NbB_2	$4d^4$	0.063	42.7
Mo_2B_5	$4d^5$	0.050	45,0
HfB_2	$5d^2$	0.100	12,0
TaB_2	$5d^3$	0.067	41.4
W_2B_5	$5d^4$	0.050	50.2

TABLE 8. Compositions, Structures, and Some Properties of Refractory Borides

Compound	Type of crystal structure	M.p., °C	Heat of formation, kcal/mole	Microhardness, kg/mm^2	Electrical resistivity, $\mu\Omega$-cm
TiB	Rhomb.	—	—	2700—2800	40
TiB$_2$	Hex.	2980	70	3370	14,4
Ti$_2$B$_5$	»	—	105	—	—
ZrB	Cub.	—	39	3500—3600	—
ZrB$_2$	Hex.	3040 ± 100	78	2252	16.6
ZrB$_{12}$	Cub.	2680	120	—	60
HfB	»	—	—	—	—
HfB$_2$	Hex.	3250 ± 100	—	2900	8,8
V$_3$B$_2$	Tetrag.	2070	—	2280	—
VB	Rhomb.	2250	—	—	35—40
V$_3$B$_4$	»	2350	—	2350	—
VB$_2$	Hex.	2400 ± 50	—	2800	79
Nb$_2$B	Tetrag.	—	—	—	—
Nb$_3$B$_2$	»	1950	—	2290	—
NbB	Rhomb.	2280	—	2195	64,5
Nb$_3$B$_4$	»	2900	—	2290	—
NbB$_2$	Hex.	3000	36	2600	34,0
Ta$_2$B	Tetrag.	—	—	—	—
Ta$_3$B$_2$	»	2120	—	2770	—
TaB	Rhomb.	2430	—	3130	100
Ta$_3$B$_4$	»	2650	—	3350	—
TaB$_2$	Hex.	3100	52	2500	37,4
CrB	Rhomb.	1750	—	1240	176
Cr$_2$B	Tetrag.	1890	—	1350	52
Cr$_5$B$_3$	Tetrag.	2000	—	—	—
CrB	Rhomb.	2050	—	1200—1300	69
Cr$_3$B$_4$	»	1950	—	1400—1500	—
CrB$_2$	Hex.	2200 ± 50	30	2100	84
CrB$_6$(?)	Tetrag.	—	—	—	—
Mo$_2$B	»	2140	22.5	2500	40
α-MoB	»	—	16.3	2350	45
β-MoB	Rhomb.	2350	—	2500	25
MoB$_2$	Hex.	2100	23	1200	45
Mo$_2$B$_5$	Rhombohedr	2100	50	2350	18
MoB$_4$	Tetrag.	—	—	—	—
W$_2$B	»	2770 ± 80	20—28	2420	—
α-WB	»	2400 ± 100	12—22	—	—
β-WB	Rhomb.	—	—	—	—
W$_2$B$_5$	Hex.	2300 ± 50	25—45	2663	43
WB$_4$	Tetrag.	—	—	—	—

In Table 8 and other tables of this type, names of crystal structures are abbreviated as follows: diam. = diamond-like; hex. = hexagonal; f.c.c. = face-centered cubic; b.c.c. = body-centered cubic; monocl. = monoclinic; orthorh. = orthorhombic; rhomb. = rhombic; rhombohedr. = rhombohedral; tetrag. = tetragonal; trig. = trigonal.

One must bear in mind, however, that the conditions of atomic interaction in borides are complicated by superposition of the additional covalent bonding between boron atoms. Hence the ratio 1/Nn can be used only to compare those borides in which the distribution of boron atoms is of the same type. The effect of this ratio on properties associated with bond character is readily seen from Table 7 [53]. As 1/Nn and the metallicity of bonding decrease, the resistivity increases.

It will be noted that the rules manifested in this case are directly related to the chemical nature of the metals forming the borides. In each triad of Table 7 the electrical resistivity increases with the degree of filling of the d shell in the transition metals by electrons.

It is curious that, as shown in Fig. 11, the melting points of these borides also depend directly on their composition and vary regularly with the position of the component elements in the periodic system. All this shows that the chemical factor plays a major part in boride formation.

More than 100 borides have been discovered. Many of them are very hard, refractory materials and have special physical properties. The compositions and main properties of some refractory borides are given in Table 8 [77].

Carbides

In the literature there are many papers and books on the investigation of carbides and description of their structure and properties [39, 40, 54, 77, 78].

Most simple carbides are interstitial phases. In this case the metal atoms form close-packed, cubic, or hexagonal structures.

As regards their physical properties, carbides are hard, refractory substances of metallic character, distinguished by their luster, high electrical conductivity, positive coefficient of thermal expansion, and ability to emit electrons. Many carbide phases form solid solutions with excess metal or carbon, and the phase diagrams contain regions of homogeneity. In this case the lattice parameter increases with carbon content. This increase causes a change in state of the metal atoms and affects the character of bonding. To judge from many data, this bonding is metallic not only between metal atoms, but also between metal and carbon atoms. Some authors [19] supposed that carbon occurred in the metallic state in interstitial phases, i.e., that it donated part of its electrons to metallic bonds, thereby becoming a positive ion. This is supported indirectly by the migration of carbon to the cathode in the electrolysis of austenite, proved in [79, 80], as well as experimental data on diffusion coefficient, electrical conductivity, and magnetic properties. The author of [19] ascribes the possibility, with regard to energy, of valence electrons passing into the d band of the metallic crystal to increase in interatomic distances on formation of interstitial phases and resulting decrease in energy of the valence electrons, which facilitates such a transition.

The increase in the degree of metallic bonding in carbides is confirmed by the increased electrical conductivity in a number of metal−carbon systems for phases corresponding to the compositions of the carbides. For instance, the conductivity in the zirconium−carbon system is maximum for the alloy of composition ZrC. There are other opinions regarding the nature of bonding in carbides. For instance, the author of [81] holds that homopolar bonds are formed between the metal and carbon atoms.

More than 100 carbides are known. Many of them are the most refractory of all compounds of metals and nonmetals. The compositions and main properties of some refractory carbides are given in Table 9.

Silicides

The composition, structure, and properties of silicides also are considered in many review articles and monographs [39, 40, 54, 55, 77, 78].

Owing to the relatively large atomic radius of silicon (see Fig. 1), silicides do not form interstitial phases. Most silicides are substitutional phases, whereas some form independent, usually complex types of crystal structures.

TABLE 9. Compositions, Structures, and Properties of Some Refractory Carbides

Composition of compounds	Type of crystal structure	M.p., °C	Heat of formation kcal/mole	Microhard- ness, kg/mm^2	Electrical re- sistivity, $\mu\Omega$-cm
TiC	Cub.	3147 ± 50	43.85	2988	52.5
ZrC	»	3530	47.7	2925	50.0
HfC	»	3890 ± 150	73.7	2913	44.0
V$_2$C	Hex.	—	11.5	—	—
VC	Cub.	2810	30.2	2094	65
Nb$_2$C	Hex.	—	—	2123	—
NbC	Cub.	3480	33.6	1961	51.1
Ta$_2$C	Hex.	3400	17.0	1714	—
TaC	Cub.	3880 ± 150	36.8	1599	42.1
Cr$_{23}$C$_6$	»	1550	25.8	1650	127
Cr$_4$C$_3$	Hex.	1665	42.52	1336	109
Cr$_3$C$_2$	Rhomb.	1895	21.01	1350	75
Mo$_2$C	Hex.	2410 ± 15	4.2	1499	71.0
MoC	»	2700	—	—	—
W$_2$C	»	2730 ± 15	7.09	3000	75.7
WC	»	2720	9.1	1781	19.2

In analogy with borides, silicon atoms in silicides form regular structures (chains, layers, skeletons) or oc-cur in isolated positions. Contrary to borides, however, some silicides contain paired silicon atoms.

The interatomic bonding in silicides is complex. There are covalent bonds between the silicon atoms; this makes possible the formation of the above-mentioned regular structures. However, silicides also contain metallic bonds, as the pronounced metallic properties of most silicides attest.

Recently several new silicides have been discovered and their properties investigated. Thus the thermo-electric properties of chromium silicides were studied in [82]. It was reported in [83] that the silicide Mn$_3$Si undergoes a phase transition at about 600°; this was confirmed by measurements of several physical properties.

Data on the compositions, structures, and main properties of some refractory silicides are given in Table 10.

Nitrides

The compositions and properties of many nitrides are described in the literature [54, 78]. Most nitrides are interstitial phases; this is easy to explain, since nitrogen, according to Fig. 1, has a smaller atomic radius ($r_N = 0.71$ A) than carbon ($r_C = 0.76$ A). Apparently nitrogen, as well as carbon, can take part in metallic bond formation; this is confirmed by ionization potentials and by measuring a number of physical properties of nitrides. In analogy with the case of carbides, the introduction of nitrogen atoms into the crystal structure of metals increases the interatomic distances and thus makes the energy conditions more favorable for the passing of valence electrons into the d band of the metallic crystal.

Nitrides resemble carbides in their properties. They have high melting points and hardness, metallic luster, and high electrical conductivity.

The compositions, structures, and properties of some refractory nitrides are given in Table 11.

The close similarity in the stoichiometric compositions of borides, carbides, and nitrides and their iso-morphism determine the character of reactions between these metallides; this will be discussed below.

A number of binary metallic compounds, containing oxygen, sulfur, phosphorus, arsenic, and some other nonmetals, also have metallic bonding. Titanium suboxides with metallic bonding were mentioned above;

TABLE 10. Compositions, Structures, and Properties of Some Refractory Silicides

Composition of compound	Type of crystal structure	M.p., °C	Heat of formation, kcal/mole	Microhardness, kg/mm^2	Electrical resistivity, $\mu\Omega$-cm
Ti_5Si_3	Hex.	2120	147,0	986	55
$TiSi$	Rhomb.	1920	39,2	1039	63
$TiSi_2$	»	1460—1540	42,9	892	16,9
Zr_2Si	Tetrag.	2220	35	—	—
Zr_5Si_3	Hex.	2250	147	—	—
Zr_3Si_2	Tetrag.	—	92	—	—
$ZrSi$	Rhomb.	2150	35.4	—	49.4
$ZrSi_2$	»	1700	—	1063	75.8
Hf_2Si	Tetrag.	—	—	—	—
Hf_5Si_3	Hex.	—	—	—	—
$HfSi$	Rhomb.	—	—	—	—
$HfSi_2$	»	1750	—	930	—
V_3Si	Cub.	1730	27,9	1430—1560	203.5
V_5Si_3	Tetrag.	—	—	—	—
V_5Si_3	Hex.	—	—	—	—
VSi_2	»	1660	—	890—960	66.5
Nb_4Si	»	2580	—	690—820	—
α-Nb_5Si_3	Tetrag.	—	—	700	—
β-Nb_5Si_3	»	—	—	—	—
Nb_5Si_3	Hex.	2400—2480	63	—	—
$NbSi_2$	»	2150	30	1050	50.4
$Ta_{4.5}Si$	»	2510	32.2	—	174.5
Ta_2Si	Tetrag.	2460	30.7	—	124
Ta_5Si_3	»	—	—	—	—
Ta_5Si_3	»	—	—	—	—
Ta_5Si_3	Hex.	—	—	—	—
$TaSi_2$	»	2200	26.2	1407	46.1
Cr_3Si	Cub.	1710±50	33.7	1005	35
Cr_5Si_3	Tetrag.	—	—	—	—
Cr_5Si_3	Hex.	—	—	—	—
$CrSi$	Cub.	1545±50	18.4	1005	129.5
$CrSi_2$	Hex.	1500±20	28.6	—	914
Mo_3Si	Cub.	2380±50	23.5	1310	21.6
Mo_5Si_3	Tetrag.	2100±50	—	—	—
Mo_5Si_3	Hex.	2320	—	—	—
$MoSi_2$	Tetrag.	2030	26.0	—	21.6
W_3Si	Cub.	—	—	770	93
W_5Si_3	Tetrag.	—	67.8	—	—
W_5Si_3	Hex.	—	46.5	—	—
WSi_2	Tetrag.	2165	22.4	1074	12.5

several phosphides [56], sulfides [57], and other compounds with metallic bonding have been reported. We shall not dwell on them. Below we shall consider metal compounds by groups, classifying them according to criteria of special physical properties. We consider the group of refractory compounds in connection with the description of borides, carbides, silicides, and nitrides. Here it is pertinent to present data on the superconductor and semiconductor groups of compounds. Recently they have become very important, and are used in new branches of engineering.

TABLE 11. Compositions, Structures, and Properties of Some Refractory Nitrides

Composition of compound	Type of crystal structure	M.p., °C	Heat of formation, kcal/mole	Microhardness, kg/mm²	Resistivity µΩ-cm
Ti_3N	Tetrag.	—	—	—	—
TiN	Cub.	3205	80.5	1994	25
ZrN	»	2080	82.2	1520	21.1
HfN	»	2982	88.24	1640	33.0
V_3N	Hex.	—	—	1900	123.0
VN	Cub.	2360	60.1	1520	85.0
Nb_2N	Hex.	2420	61.1	1720	142.0
$NbN_{0.79}$	Tetrag.	—	—	—	—
$NbN_{0.98}$	Hex.	—	—	—	—
$NbN_{0.94}$	Cub.	—	—	—	—
NbN	Hex.	2300	56.8	1396	78.0
Ta_2N	»	2050	64.7	1220	263.0
TaN	»	3087±50	60.0	1060	128.0
Cr_2N	Hex.	1650	25.2	1571	76
CrN	Cub.	Dissociates at 1500°	—	1093	640
Mo_3N	Tetrag.	Dissociates at 600°	—	—	—
Mo_2N	Cub.	—	16.6	630	19.8
MoN	Hex.	—	—	—	—
W_2N	Cub.	—	17.2	—	—
WN	Hex.	Decomposes at 600°	—	—	—

Superconductor Compounds

It was established as a result of many investigations [84-96] that numerous elements (metals) have superconductor properties. Such metals lose their electrical resistivity at temperatures near absolute zero. The resistivity of a semiconductor is about 10^{-14} times that of pure copper. The temperature in degrees Kelvin, at which transition to the superconductive state occurs, is called the critical temperature and denoted by T_C.

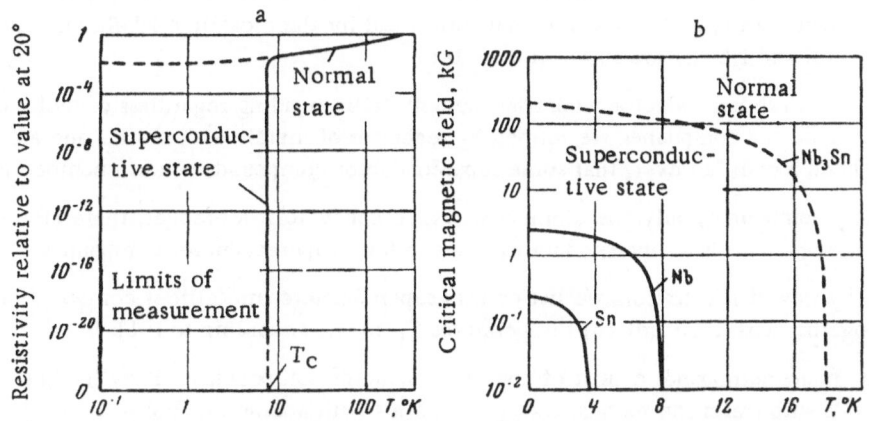

Fig. 13. Diagrams of transition to superconductive state in niobium metal (a) and temperature dependence of critical magnetic field for superconductive metals—tin, niobium, and metallide Nb_3Sn (b).

A diagram of the transition of niobium metal to the superconductive state is shown in Fig. 13a. As is evident, the electrical resistivity decreases gradually at first with decreasing temperature and then abruptly falls nearly to zero at a certain temperature. For niobium, this temperature is 9.17°K. It is the critical temperature for transition of niobium to the superconductive state.

A second important characteristic of superconductive materials is the fact that they lose their superconductivity in strong magnetic fields or when an intense (critical) current is passed through the superconductor.

In this connection, each substance also has its own critical magnetic field and current density, at which its superconductivity vanishes. The critical field intensity H_0 varies with temperature, increasing from zero at the critical temperature toward a maximum as the temperature approaches absolute zero. The corresponding curves of critical magnetic field versus temperature for tin and niobium are shown in Fig. 13b. They determine the critical magnetic fields for these two metals and their variations at temperatures from 0 to 4-8°K. Curves of the critical magnetic field for the metallide Nb_3Sn are shown here for comparison.

Of the 102 elements, only 24 have been found to possess superconductor properties up to now. They are all metals. They are listed in Table 12, together with their critical temperatures (calculated and experimental) and critical magnetic fields. To this main table are added four elements which exhibit superconductivity in thin films prepared below 10°K, as well as bismuth, for which T_c varies with the external pressure.

As is evident from Table 12, all superconductive metals have relatively low T_C values; niobium has an exceptionally high T_C value (9.17°K), whereas T_C for most elements is below 1-4°K. The problem of practical application of the superconductor properties of materials in engineering would be exceedingly limited if it were not for the possibility of discovering new superconductive materials among solid solutions, eutectic mixtures of alloys, and metallic compounds. In this connection, a very important part was played by theoretical investigations in physics, physicochemical analysis, and metal chemistry. Investigations of reactions of metals and equilibrium diagrams of many metallic systems revealed extensive possibilities of finding and developing numerous new compositions of alloys and compounds with pronounced superconductor properties.

Hundreds of new superconductive alloys and compounds have been discovered in the last 15-20 years as a result of these studies [84-96]. Many of them have higher critical temperatures of transition to the superconductive state than pure metals, as well as higher critical magnetic fields and current densities. The highest critical temperatures were found in niobium and vanadium compounds. The highest T_C values for metallides discovered in recent years are 17.5-18.5°K (Nb_3Al, Nb_3Sn, V_3Ga, V_3Si), i.e., about twice the value for niobium, which is the highest among pure metals.

For comparison of the maximum critical parameters of superconductive compounds and alloys, curves of critical magnetic field versus temperature are plotted for the metallides Nb_3Sn and V_3Ga in Fig. 14a, whereas curves of critical current density at 4.2°K versus magnetic field for the metallide Nb_3Sn and alloys of the systems Pb−Bi, Mo−Re, and Nb−Zr are shown in Fig. 14b.

All compounds with superconductor properties have metallic bonding regardless of their component elements. Many superconductor compounds are formed by reactions of metals with metals and are typical intermetallides. It has been found, however, that some superconductor compounds contain nonmetals (B, C, Si, etc.).

In many binary compounds, only one element is a superconductor. Moreover, there are many cases where two elements, neither of which is a superconductor, react to form superconductor compounds.

In Table 13, by way of illustration, we list certain superconductor metallides composed of nonsuperconductor elements, together with their critical temperatures T_C in °K, according to [91].

As is evident, many compounds consist of nonsuperconductor components. They include compounds of electropositive metals with transition metals, transition metals with nonmetals, etc.

By 1961 the number of known superconductor materials (including pure metals) was about 500; today, according to [91], it is 900. They are mainly metal compounds and solid solutions. The search for new alloys

TABLE 12. Properties of Superconductive Elements

Element	T_c, °K		H_0, Oe	
	calculated	found	calculated	found
Al	1.183	1.196	104	99
Cd	0.54	0.56	29	30
Ga	1.087	1.091	59.4	51
Hg (α)	—	4.153	—	412
Hg (β)	—	3.949	—	339
In	3.396	3.4035	248	293
Ir	—	0.14	—	19
La (α)	4.80	5.0	—	—
La (β)	5.91	6.3	—	1600
Mo (high purity)		0.92	—	98
Nb	9.17	9.13	1944	1980
Os	—	0.655	—	65
Pb	7.23	7.193	—	803
Re	1.699	1.698	188	198
Ru	—	0.49	—	66
Sn	3.722	3.722	303	309
Ta	4.39	4.483	780	830
Tc	—	8.22	—	—
Th	—	1.368	131	162
Ti	—	0.39	—	100
Tl	2.36	2.39	—	171
U (α)	—	0.68	—	—
U (γ)	—	1.80 (extrap.)	—	—
V	5.03	5.30	1310	1020
Zn	0.852	0.875	51.8	53
Zr	—	0.546	—	47

Note. Thin films prepared at T < 10°K have the following critical temperatures (in °K): Be, 6.0-8.4; Bi, 6.0; Ge, 8.4; In, 3.95-4.25; Sn, 4.6-4.7.

The critical temperature of bismuth varies with pressure, being 3.916 °K at P = 25,000 atm and 7.25 °K at P = 26,000-28,400 atm.

and compounds with superconductor properties is far from over, and further investigations of reactions of metals with metals and nonmetals in simple and many-component systems will uncover many new superconductor materials with hitherto unheard-of properties. The number of such compounds may be increased also as a result of reactions between nonsuperconductor elements.

Until now, questions of the theory of formation of superconductor compounds have scarcely been developed. They are considered in the three review articles cited above [89-91]. It should be recognized that we have no clear ideas as yet regarding the actual conditions for the appearance of superconductor compounds or alloys in one system and their absence in another.

Based on existing experimental data, one can say that compounds of transition metals in Groups V and VI have the highest T_c values; in most such compounds and solid solutions containing them, the ratio of the number of electrons to the number of atoms in the crystal structure (Matthias's rule) is 1-10 valence electrons

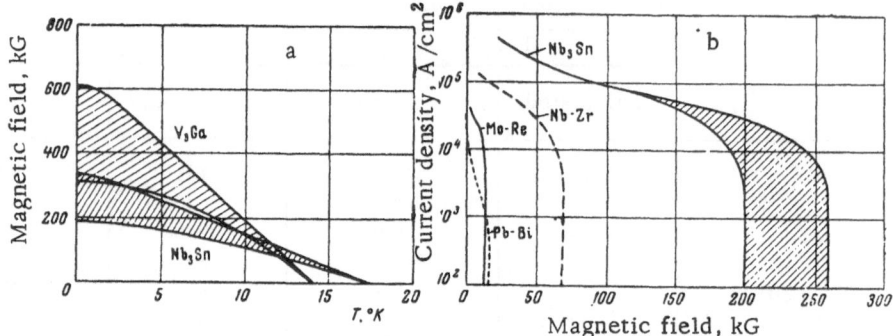

Fig. 14. Maximum values of critical magnetic field versus temperature for compounds Nb_2Sn and V_3Ga (a) and critical current density versus magnetic field at 4.2 °K for alloys Pb—Bi, Mo—Re, Nb—Zr, and metallide Nb_3Sn (b).

TABLE 13. Superconductor Compounds Formed by Nonsuperconductor Elements

Composition of compound	T_C, °K	Composition of compound	T_C, °K
$SrRh_2$	6.20	W_2C	2.74
$BaRh_2$	6.00	Pd_xSe_y	2.50
$CaIr_2$	6.15	PdTe	2.30
$SrIr_2$	5.70	W_3P	2.26
Rh_xSr_y	6.00	Pd_2Se	2.20
Rh_9S_8	5.80	YIr_2	2.18
W_3Ir	3.82	CuS	1.62
WRh	3.37	$CoSi_2$	1.40
W_2B	3.10	Rh_5P_4	1.22
W_3Si_2	2.84	ScIr	1.03

per atom [85, 90]. Subsequent investigations in this direction [92] narrowed the range of these ratios to 4.6-4.7 instead of Matthias's 1-10. However, all these values are empirical; they do not have an adequate theoretical basis as yet, and there are many exceptions.

In superconductive materials, considerable importance is attached to the crystal structure of alloys and compounds and the arrangement of atoms in them, and also to finding the relation between structure type and superconductor properties [89-92]. It was found that the most favorable types of crystal structure for superconductor compounds are the cubic β-W (A15), hexagonal $MgZn_2$ (C14), and cubic $MgCu_2$ (C15), the structures of various compounds of the Laves-phase type, and the cubic α-Mn (A12) structure. Compounds with the β-W, NaCl, $MgCu_2$, and $MgZn_2$ structures have the highest critical temperatures [93]. Superconductor compounds based on niobium are described in detail in several recent papers [94, 95].

Out of more than 900 superconductor substances having critical temperatures from ~0.07 to 18.5 °K, the most important compounds and some alloys based on them, having T_C values above 0.3 °K, are given in Table 14 [90, 91]. Many of these compounds are promising for practical application.

Table 14 lists the compositions, types of crystal structure, and critical temperatures T_C of these materials. It also includes all metallides formed by transition metals with other elements. Besides binary superconductor compounds of stoichiometric composition, the table includes some ternary compounds ($SnNbV_2$, $SnNbTa_2$,

PbAsBi, $SnNb_2Ta$, etc.), and a number of binary and ternary alloys having the listed atomic ratios $Nb_{0-0.15}$ $Ti_{1-0.85}$; $Ta_{1-0.7}Ti_{0-0.3}$, etc.), which are solid solutions of metals or metallides

Based on a consideration of superconductivity data for all materials in the table, the author of [91] concludes that in some cases T_C varies considerably for a given substance; for instance, niobium nitride NbN has three T_C values: ≤ 5.10, 5.58, and 15.60 °K. Here T_C obviously depends on the purity of the compounds obtained.

The properties and structures of the superconductor compounds given in Table 14, i.e., their similarities and differences, especially with regard to type of crystal structure, are very important for investigating interactions of these compounds. The development of investigations along these lines should result in the establishment of some general rules governing the variation of properties with composition, as well as new superconductor materials with new properties. Some investigations of new systems of superconductor compounds will be described below.

Semiconductor Compounds

As a new class of materials having special physical properties, semiconductors have recently received special attention owing to their extensive prospects for technical application [62, 65, 97-100]. The general name semiconductors emphasizes the specific electrical properties of these materials. In the series of solids, whose extreme positions are the metallic and ionic types of chemical bonding, semiconductors occupy an intermediate position. In the case of electronic conduction, which is characteristic of metals and many of their compounds (see above), valence electrons migrate freely in an external electric field. Therefore metals have high electrical conductivity. In ionic conduction, positive or negative ions migrate. Compounds with ionic bonding do not conduct current; they are insulators.

Current is carried in semiconductors, as in metals, by the migration of valence electrons; they differ from metals only by their lower electrical conductivity. This fact brings semiconductors close to metals in properties, and semiconductor compounds close to metallic compounds.

With respect to electrical conductivity, semiconductors are intermediate between conductors and insulators, as will be seen below. Highly conductive metals and some of their compounds have conductivities $\sigma = 1/\rho = 10^4 - 10^6 \ \Omega^{-1} \cdot cm^{-1}$; for insulators this quantity is usually less than $10^{-10} \ \Omega^{-1} \cdot cm^{-1}$. The conductivity of semiconductors varies in a broad interval, from 10^3 to $10^{-10} \ \Omega^{-1} \cdot cm^{-1}$, between the conductivities of metals and insulators.

A principal characteristic of semiconductors is the high sensitivity of their properties, particularly conductivity, to minute traces of impurities in the substance and to external factors — temperature, pressure, illumination, etc. These factors strongly affect the properties of semiconductors; the latter differ from metals in this respect.

For instance, the electrical resistivity of metals and alloys increases with rising temperature, whereas that of semiconductor compounds, on the contrary, decreases. This is due to the different behavior of the outer electrons in metal and semiconductor atoms.

The most important characteristic of semiconductors is the activation energy of intrinsic conductivity or, as it is called, the width of the forbidden band, measured in electron volts (ΔE, eV). In describing the properties of semiconductors, this is the main quantity. With respect to the character of conductivity, these compounds are classified as electronic and ionic semiconductors [97]. Another important property of semiconductors is the type of conductivity. In the case of electronic, or n-type, conductivity the current is carried mainly by free electrons, which do not form valence bonds.

In hole, or p-type, conductivity the current is carried by "holes" resulting from a deficiency of electrons in the valence bonds. The type of conductivity in semiconductors is easily determined by the sign of the Hall coefficient or by indices of thermo-emf.

TABLE 14. Compositions, Structures, and Property T_c of Superconductor Compounds

Compound	Type of crystal structure	$T_c, °K$	Compound	Type of crystal structure	$T_c, °K$
Zr_2Al_3	Hex. $(MgZn_2)$	<0.3	Th_7Rh_3	Hex. (Fe_3Th_7)	2.15
Cr_3Ir	Cub. $(β\text{-}W)$	0.45	Cr_3Ru	Cub. $(β\text{-}W)$	2.15
Ti_3Pt	Cub. $(β\text{-}W)$	0.58	ZrW_2	Cub. (Cu_2Mg)	2.16
Sb_2Au	Cub. (FeS_2)	0.58	YIr_2	Cub. (Cu_2Mg)	2.18
Mo_3Al	Cub. $(β\text{-}W)$	0.58	Pd_5Se	—	2.2
$TiCO$	Cub. $(β\text{-}W)$	0.71	$CuBi$	—	2.2
Zr_3Al	Cub. (Cu_3Au)	0.73	$NaBi$	Tetrag. $(CuAu)$	2.25
$ThAl_3$	Hex. (Ni_3Sn)	0.75	Th_2Ag	Tetrag. $(CuAl_2)$	2.26
Zr_3Pb	Cub. $(β\text{-}W)$	0.76	W_3P	—	2.26
V_3Sb	Cub. $(β\text{-}W)$	0.80	U_6Co	Tetrag. (MnU_6)	2.29
Mg_2Al_3	Cub.	0.84	$PdTe$	Rhombohedr.(Mo_2B_5)	2.3
$PdSi$	Orthorh. (MnP)	0.93	Pd_4Se_2	—	2.3
$MoIr$	Hex. (Mg)	<1.0	U_6Mn	Tetrag. (MnU_6)	2.32
$TaPt$	Tetrag. $(CrFe)$	1.0	$AuSn_4$	Orthorh. $(PtSn_4)$	2.38
$ScIr_2$	Cub. (Cu_2Mg)	1.3	$NbPt$	Tetrag. $(CrFe)$	2.4
Nb_3Os	Cub. $(β\text{-}W)$	1.05	$PtBi$	Hex. $(NiAs)$	2.4
$TiRu$	Cub. $(CsCl)$	1.07	$ThSi_2 (β)$	—	2.41
KHg_2	—	1.2	$AlIn_2$	Tetrag. $(CuAl_2)$	2.30—2.46
$AuZn_3$	—	1.21	$LiBi (α)$	—	2.47
$CoSi_2$	Cub. $(CaFe_2)$	1.22	$AuSn_2$	—	2.48
$PdSb_2$	Cub. (FeS_2)	1.25	Nb_3Rh	Cub. $(β\text{-}W)$	2.5
Mo_3Si	Cub. $(β\text{-}W)$	1.3	$LaSi_2$	Tetrag. (Si_2Th)	~2.5
$ReTa$	Tetrag. $(CrFe)$	1.3	$NbOs_2$	Cub. $(α\text{-}Mn)$	2.52
$ScGe_2$	—	1.30—1.31	Ru_7B_3	Hex. (Fe_3Th_7)	2.58
$NbOs$	Tetrag. $(CrFe)$	1.40	$NbSn_2$	—	2.6
Mo_3Ge	Cub. $(β\text{-}W)$	1.43	$BeAu$	Cub. $(FeSi)$	2.64
$PdSb$	Hex. $(NiAs)$	1.5	$RhPb_2$	Tetrag. $(CuAl_2)$	2.66
$RhTe_2$	Cub. (FeS_2)	1.51	$AgTl$	—	2.67
Th_7Os_3	Hex. (Fe_3Th_7)	1.51	$HfOs_2$	Hex. $(MgZn_2)$	2.69
YRu_2	Hex. $(MgZn_2)$	1.52	$TlIn$	—	2.7
Zr_2Ni	—	1.52	$ZrRh$	—	2.7
Th_7Ir_3	Hex. (Fe_3Th_7)	1.52	$RhBi_4 (γ)$	Hex., complex	2.7
$PdTe$ (single crystal)	Trig. (CdI_2)	1.53	W_2C	—	2.74
			$MgTl$	—	2.75
YPt_2	Cub. (Cu_2Mg)	1.57	Mo_2C	—	2.78
Ta_2Ge	—	1.6	$PtPb_4$	Tetrag. $(CuAl_2)$	2.8
$NaHg_2$	Hex.	1.62	Re_2B	—	2.8
CuS	Hex. (CuS)	1.62	V_3Pt	Cub. $(β\text{-}W)$	2.83
$LaRu_2$	Cub. (Cu_2Mg)	1.63	W_3Si_2	—	2.84
Nb_3Ir	Cub. $(β\text{-}W)$	1.7	$PdPb_2$	Tetrag. $(CuAl_2)$	2.95
$LiHg_3$	Hex.	1.7	$ZrPt$	Hex. (Mg)	3.0
UCo	Cub. $(CsCl)$	1.7	$ZrOs_2$	Hex. $(MgZn_2)$	3.0
Pd_2As	—	1.7	W_3Os	—	2.21—3.02
Zr_3Sb_2	—	1.74	$NaHg_4$	—	3.05
$HgCd$	Tetrag.	1.77	Th_2Au	Tetrag. $(CuAl_2)$	3.08
Th_7Co_3	Hex. (Fe_3Th_7)	1.83	W_2B	Same	3.1
YRe_2	—	1.83	Ta_2B	» »	3.12
Au_2Bi	Cub. (Cu_2Mg)	1.84	$ThSi_2 (α)$	—	3.16
$ZrRu_2$	Hex. $(MgZn_2)$	1.84	KHg_2	—	3.18
Nb_3Os_2	Tetrag. $(CrFe)$	1.78—1.85	$RhBi_3$	$NiBi_3$	3.2
Th_7Fe_3	Hex. (Fe_3Th_7)	1.86	ZrB	Cub.	2.8—3.2
Pd_5As_3	—	1.9	KHg_4	—	3.27
Nb_2Ge	—	1.9	Cr_3Ru	Cub. $(β\text{-}W)$	3.3
$AuTl$	—	1.92	Ta_3C	—	3.3
$TaOs$	Cub. $(α\text{-}Mn)$	1.95	WRh	Hex. (Mg)	2.64—3.37
$MoRh$	Hex. (Mg)	1.97	KHg_3	—	3.42
$NbRe$	Tetrag. $(CrFe)$	2.0	Ti_2Co	—	3.44
$NbPd$	Same	2.0	$SnBi$	—	3.48
$TaRh$	» »	2.0	Th_2Cu	Tetrag. $(CuAl_2)$	3.49
RuC	—	2.0	$ThRu_2$	Cub. (Cu_2Mg)	3.56
$CaBi_3$	—	2.0	KBi_3	Cub. (Cu_3Mg)	3.58
Cr_2Ru	Tetrag. $(CrFe)$	2.02	KBi	—	3.6
$RhBi$	Hex. $(NiAs)$	2.06	$CdSn$	—	3.65
$PtSb$	Hex. $(NiAs)$	2.1	$PdBi$	Orthorh.	3.7
Rh_5Ge_3	—	2.12	$AuSn$	—	3.7

STRUCTURE AND COMPOSITIONS OF METALLIDES

33

TABLE 14 (Continued)

Compound	Type of crystal structure	T_c, °K	Compound	Type of crystal structure	T_c, °K
AgSn	—	3.7	Ti_xV_y	—	5.3—7.25
Sn_2Sb_2	—	3.8	$Ti_{15-8}V_{85-2}$	Cub. (β-W)	3.5—7.3
V_2Sn	Cub. (β-W)	3.8	$Nb_3Sn_2V_3$	Cub. (β-W)	7.4
YGe_2	Tetrag. (Si_2Th)	3.8	$Nb_{0.2-3.6}$	—	1.7—7.5
SnBi	—	3.81	VN	Cub. (NaCl)	7.5
HgIn	—	3.81	$Ti_{0-85}V_{1-15}$	Cub. (β-W)	2.3—7.5
SnO	—	3.81	$NbO_{10}N_{66}$	Tetrag.	8.1
U_6Fe	Tetrag. (MnU_6)	3.86	VN	Cub. (NaCl)	8.20
Pd_2Bi	—	4.0	NbB	Orthorh.	8.25
$NiBi_3$	Orthorh.	4.06	ZrV_2	Cub. $(MgCu_2)$	8.8
Tl_2Pb	—	3.75—4.09	ZrN	Cub. (NaCl)	6.9
NbRh	Tetrag. (CrFe)	4.1	Re_2W	Cub. (α-Mn) -- BC	9.0
ZrIr	Cub. (Cu_2Mg)	4.1	AsBiPbSb	—	9.0
SnAs	—	4.1	AsBiPb	—	9.0
HgSn	—	4.2	ZrN	Cub. (NaCl)	9.05
NiBi	Hex. (NiAs)	4.25	Nb_3Pt	Tetrag.	9.2
$RbBi_2$	Tetrag.	4.25	Nb_3In	Cub. (β-W)	9.2
TaSi	Hex.	4.25—5.38	MoC	—	9.26
OsW	Tetrag.	4.4	Nb_3Ti	Cub.	9.3
$AuPb_2$	Tetrag. $(CuAl_2)$	4.42	$Nb_{02-5}Ti_y$	—	5.8—9.5
Zr_5Pb_3	Hex. (Mn_5Si_3)	4.6	$MoGa_2$	—	9.5
$ScOs_2$	Hex. $(MgZn_2)$	4.6	$V_{2.48}Ga$	Cub. (β-W)	9.6
YOs_2	Same	4.7	Mo_3Re	—	9.8
IrGe	Orthorh. (MnP)	4.7	Nb_3SnV	Cub. (β-W)	9.8
Mo_2B	Tetrag. $(CuAl_2)$	4.74	ZrN	Cub. (NaCl) -- FC	10.7
$CsBi_2$	Cub. (Cu_2Mg)	4.75	$Zr_{0.25}Nb_{0.75}$	—	10.8
$HfRe_2$	Hex. $(MgZn_2)$	4.8	Nb_3Zr	—	10.8
TiN	Cub. (NaCl)	4.86	$NbSnTa_2$	Cub. (β-W)	10.8
$CeRu_2$	Cub. (Cu_2Mg)	4.9	TaC	—	11.0
Mo_2N	Cub.	5.0	MoN	Hex. ord.	12.0
WRe	Tetrag.	5.2	Si $(V_{0.9}Nb_{0.1})_3$	Cub. (β-W)	12.8
W_2C	Cub.	5.2	Nb_3Ga	Cub. (β-W)	12.5—13.2
Ti_3Ir	Cub. $(MgCu_2)$	5.4	NbC	—	14.0
$Nb_{0.15}Ti_{0.85}$	Cub.	0.6—5.5 (7.9 annld.)	$Nb_{2.5}SnV_{0.5}$	Cub. (β-W)	14.2
$NbSnV_2$	Cub. (β-W)	5.5	V_3Si (1,3%Fe, Mn)	Cub. (β-W)	14.4
TiN	Cub. (NaCl)	5.6	Nb_3Ga	Cub. (β-W)	14.5
Ti_3Sb	Cub. (β-W)	5.8	NbN	Hex.	15.6
$OsAl_2$	—	5.9	V_3Si (0,4%Fe, Mn)	Cub. (β-W)	16.3
$N_{0.58}Nb_{0.02}$	Tetrag.	6.0	$Nb_3Al_2Sn_5$	Cub. (β-W)	16.3
Mo	Tetrag.	6.0	$SnNb_3Ta$	Cub. (β-W)	16.4
$ZrRe_2$	Hex. $(MgZn_2)$	6.0	V_3Ga	Cub. (β-W)	16.8
Re_3W_2	»	6.0	V_3Si (0,25%Fe, Mn)	Cub. (β-W)	17.0
Ta_3Sn	Cub. (β-W)	6.0	V_3Si	Cub. (β-W)	17.0
NbC	Cub. (NaCl)	6.0	Nb_3Al	Cub. (β-W)	17.1
NbSnTaV	Cub. (β-W)	6.2	$Nb_{2.5}SnTa_3$	Cub. (β-W)	17.6
$Nb_{0-25}Ti_4$	Hex.	1.6—6.3	$Nb_{2.75}SnTa_{0.25}$	Cub. (β-W)	17.8
Ti_9V	—	6.3	Nb_3Al	Cub. (β-W)	17.5—18.0
$Ta_{1-7}Ti_{0-3}$	—	4.3—6.5	$Nb_{3x}SnTa_3$ (1—x)	Cub. (β-W)	6.0—18.0
PbSb	—	6.6	Nb_3Sn	Cub. (β-W)	18.05
$Ti_{0.85}V_{0.15}$	—	0.6—4.4 (6.6 annld.)	Nb_xSn_y	—	17.25—18.18
$Ti_{0-5}V_{1-5}$	Cub. (β-W)	5.3—6.7	$Nb_3Sn_{1.7}In_{0.3}$	Cub. (β-W)	18.0—18.19
$Fe_xTi_6V_4$	—	6.8	$Nb_{76}Sn_{24}$	Cub. (β-W)	18.2
Mo_3P	—	7.0	$Nb_3Sn_{1-x}Bi_x$	—	18.0—18.2
Nb_4N_3	Tetrag.	7.2	$Nb_{0.8}Sn_{0.2}$	Cub. (β-W)	18.5
LiPb	—	7.2			

Note. Arbitrary notation for structure types: cub. (Cu) — A1; cub. (W) — A2; cub. (α-Mn) — A12; cub. (β-W) — A15; hex. (Mg) — A3; diam. — A4; cub. (NaCl) — B1; cub. (CaF_2) — C1; hex. $(MgZn_2)$ — C_{14}; cub. $(MgCu_2)$ — C15.

TABLE 15. Structures and Properties of Semiconductor Elements

Element	Melting point, °C	Type of crystal structure	Forbidden-band width ΔE, eV
Boron	2050	Hex.	1.55; 1.1*
Carbon	~3820	Hex.	0.0
	(P ± 120,000 atm)	Cub. (diamond)	6.00
Silicon	1420	Cub. (diamond)	1.12
Germanium	940	Cub. (diamond)	0.75
Tin (gray)	—	Cub. (diamond)	0.08
Sulfur	112	Rhombohedr.	2.50
Selenium	220	Hex.	2.30
Tellurium	456	Hex.	0.33
Phosphorus (black)	44	Rhomb.	0.33
Arsenic	500	Rhombohedr.	1.20
Antimony	630	Rhombohedr.	0.12
Iodine	113	Hex.	1.25

* 1.55 in single crystals; 1.1 in polycrystalline samples.

As in the case of superconductors, the number of chemical elements having semiconductor properties proved to be very limited. Nature was poor in this respect. Of the total number of 102 elements, only 12 have semiconductor properties. These are all located in B subgroups; of them, boron lies in Group III; carbon, silicon, germanium, and tin (α-form) in Group IV; phosphorus, arsenic, and antimony in Group V; sulfur, selenium, and tellurium in Group VI; and iodine in Group VII. None of these elements is a superconductor. They all have complex crystal structures and covalent bonding.

The semiconductor character of the above elements is due to the predominance of this type of bonding.

Some properties of these semiconductor elements are given in Table 15.

Of all these elements, germanium and silicon have the most distinctive semiconductor properties. Today they have found extensive practical application in very pure form. Antimony, arsenic, tellurium, selenium, etc., are very interesting as semiconductors, not so much in pure form as in the form of alloys and compounds with other elements.

The paucity of semiconductor elements is compensated to a considerable degree, as in the case of superconductor materials, by the possibility of forming a large number of metal compounds and alloys having semiconductor properties. In this field of new materials, hundreds of alloys and compounds with very interesting and practically important semiconductor properties have been discovered through the efforts of chemists and physicists [100-105].

The particularly fruitful investigations of Kurnakov, founder of physicochemical analysis, and his students in the field of equilibrium diagrams of metallic systems containing semiconductor elements should be noted [8, 14, 15]. These studies began with his investigation of intermetallic compounds [8], where questions of the formation of many metallic compounds, including some with semiconductor properties unknown at that time, were considered from a general chemical viewpoint.

In connection with this, one should recall a number of investigations of magnesium alloys with tin, lead, and bismuth, conducted by N. I. Stepanov at the beginning of this century [106, 107]. As a result of them, he discovered (in 1905) compounds in these systems, whose atomic ratios conform exactly to the rules of valence: Mg_2Pb, Mg_2Sn, and Mg_3Bi_2. These substances, which later were called valence compounds, are daltonides and have singular points in the property — composition diagrams of these systems. They have high electrical resistivity and, except for Mg_2Pb, are semiconductor compounds. Later the number of such magnesium compounds with elements of Groups IVB and VB was increased by the discovery of analogous metallides: Mg_2Si and Mg_2Ge, Mg_3Sb_2 and Mg_3As_2, etc. [43]. They all proved to be semiconductor compounds.

Proceeding from the general term metallides for such compounds, N. I. Stepanov proposed that these compounds be called magnesium plumbides, stannides, and bismuthides [106]. These names indicate the electropositivity of magnesium and electronegative properties of the lead, tin, and bismuth reacting with it. Such names were widely used by Kurnakov and his co-workers (S. F. Zhemchuzhnyi, A. N. Kuznetsov, N. S. Konstantinov, et al.) in announcing new compounds of sodium with cadmium (cadmides), iron and cadmium with antimony (iron and cadmium antimonides), etc. [8]. Many other compounds of metals with metals and nonmetals were named according to this principle which later proved to be semiconductor compounds.

Such names are widely used today in the field of semiconductor compounds. Here one could list compounds with such names as aluminides, arsenides, oxides, selenides, silicides, sulfides, tellurides, phosphides, and many others. Within the meaning of the name adopted by us — metallides — we regard all semiconductor compounds as metallides.

The chemistry of semiconductor compounds has been developed extensively by investigators of various countries. Numerous monographs and review articles are devoted to it [62, 65, 101-105]. The most detailed review of semiconductor compounds with the diamond structure is given in [105].

The theory of the formation and properties of semiconductor compounds need not be considered here. We shall give only a brief review of some of the ways in which the formation of semiconductor compounds depends on the position of their component elements in the periodic system, and a description of the compositions and main properties of such compounds.

First of all, it should be noted that one of the component elements of a semiconductor compound is more electropositive than the component semiconductor element. In semiconductor compounds with the general formula A_xB_y, as a rule, element A is a metal of Group I, II, or III, and the subscripts x and y represent group numbers (e.g., A_I, A_{II}, A_{III} are metals of the first, second, and third groups, etc.). This element A_x may be one of many metals more electropositive in the given compound than the element B_y. The latter may be any semiconductor element, i.e., any element of Groups IVB, VB, VIB, VIIB, and possibly IIIB. In accordance with this, these elements are denoted by B_{IV}, B_V, etc., in the formulas of semiconductor compounds.

Thus, of the metal compounds with the general formula A_xB_y, those formed between a large number of metals in subgroups A and B and semiconductor elements of Groups III-VII, subgroup B, are semiconductor compounds.

In this case it is easy to determine how compounds with different types of chemical bonding, depending on the group and subgroup of each component element, are formed. In many instances one can follow the transition from purely metallic bonding to the covalent type characteristic of semiconductor compounds and to ionic bonding in them. The character of chemical bonding and its gradual transition from metallic to ionic are determined by the relative difference in electronegativity of the elements and the electronic structure of the atoms. It can be shown that metal compounds with the general formula A_IB_{II} have metallic bonding, compounds of the types $A_{II}B_{III}$, $A_{II}B_{IV}$, $A_{II}B_V$, $A_{II}B_{VI}$, and $A_{III}B_V$ have covalent bonding and are semiconductors, and compounds of the types A_IB_{VII} have ionic bonding.

Empirical values of electronegativity differences may be used to distinguish between semiconductor and ionic compounds. When the electronegativity difference is about 2 or less, the compounds are semiconductors; when this difference is larger than 2, they are insulators and have ionic bonding [108]. The forbidden-band width also may be used for this purpose: If the width is less than 2 eV, the compound is a semiconductor; if the width is more than 2 eV, it is an insulator. Intermediate types of bonding are observed between these limiting values.

As will be shown below, this order of changes in semiconductor properties obtains in the case of many binary and ternary metal compounds.

TABLE 16. Data on the Compositions and Certain Properties of Semiconductor Compounds

Composition of compound	Melting point, °C	Type of crystal structure	Forbidden-band width ΔE, eV
$A_{II}B_{IV}$ Compounds			
Mg_2Si	1102	Cub. (CaFe)	0.77
Mg_2Ge	1070	Same	0.55
Mg_2Sn	778	»	0.25
Mg_2Pb	550	»	0.0
Ca_2Si	920	Tetrag.	1.90
Ca_2Ge	—	Tetrag.	—
Ca_2Sn	1122	Tetrag.	0.90
Ca_2Pb	1110	Tetrag.	0.50
$A_{II}B_{VI}$ Compounds			
ZnS	1850	Cub. (ZnS)	3.60
$ZnSe$	~1500	Same	2.68
$ZnTe$	1240	»	2.25
CdS	1750 (P = 100 atm)	»	2.42
$CdSe$	1350	»	1.74
$CdTe$	1105	»	1.50
HgS	1750 (1200мм)	»	2.50
$HgSe$	690	»	—
$HgTe$	600	»	~0.01
$SrSe$	—	Cub. (NaCl)	~2.00
$SrTe$	—	Cub. (NaCl)	~2.00
$A_{III}B_{V}$ Compounds			
AlP	Decomposes above 1000°	Cub. (ZnS)	3.00
$AlAs$	1600	Same	2.10
$AlSb$	1060	»	1.55
GaP	~1850	»	2.35
$GaAs$	1280	»	1.35
$GaSb$	728	»	0.70
InP	1055	»	1.30
$InAs$	942	»	0.33
$InSb$	525	»	0,17
$AlIn$	High	Hex. wurtzite	—
GaN	High	Hex. wurtzite	—
InN	High	Hex. wurtzite	—
$InBi$	110	Tetrag.	Metallic
$TlBi$	230	Cub. (CsCl)	—
Other sulfides, selenides, and tellurides			
PbS	1114	Cub. (NaCl)	0.41
$PbSe$	1065	Same	0.29
$PbTe$	905	»	0.32
$SnSe$	861	—	1.30
$SnTe$	780	Cub. (NaCl)	—
As_2Se_3	360	Rhombohedr.	—

Composition of compound	Melting point, °C	Type of crystal structure	Forbidden-band width ΔE, eV
As_2Te_3	362	—	1.20
Sb_2S_3	550	Rhomb.	1.70
Sb_2Se_3	611	Rhomb.	1.20
Sb_2Te_3	620	—	0.35
Bi_2S_3	850	Rhomb.	1.20
Bi_2Se_3	710	Rhombohedr.	0.28
Bi_2Te_3	580	Rhombohedr.	0.15
$InSe$	660	—	1.20
$InTe$	696	—	—
In_2Te_3	667	—	—
Tl_2S	443	—	~1.00
Tl_2S_3	310	Tetrag.	—
FeS_2	—	Cub. (FeS$_2$)	1.20
$FeSe_2$	—	Cub. (FeS$_2$)	—
$MnSe$	—	Cub. (NaCl)	2.50
$MnSe_2$	—	Cub. (FeS$_2$)	0.15
MoS_2	1175	Cub.	~1.00
Ag_2S	850	Cub. (CuF$_2$)	~1.00
Ag_2Se	~1000	—	~0.075
Ag_2Te	958	Monocl.	0.17
GeS			1.80
Other arsenides, antimonides, and bismuthides			
Mg_3As_2	800	Cub.	—
Mg_3Sb_2	961	Hex.	0.82
Mg_3Bi_2	715	Hex.	—
$CdSb$	456	Rhomb.	0.48
$ZnSb$	544	Rhomb.	0.56
Zn_3As_2	1015	Tetrag.	1.00
Cd_3P_2	—	Tetrag.	0.60
Cd_3As_2	721	Tetrag.	0.50
$CsSb$	—	—	0.80
Cs_3Sb	—	—	0.80
Cs_3Bi	—	—	0.60
Zn_3Sb_2	568	—	—
Mn_2Sb	950	Tetrag.	—
$FeAs$	1050	Rhomb.	—
$MnAs$	935	Rhomb.	—
Mn_2As	1029	—	—
Ternary semiconductor compounds [105,110]			
$CuFeS_2$	—	Cub. chalcopyrite	0.53
$CuInS_2$	950	Same	1.20
$CuInSe$	—	»	0.92
$CuInTe$	—	»	0.95
$AgInS_2$	850	»	1.90
$AgInSe_2$	—	»	1.18
$AgInTe_2$	—	»	0.96

TABLE 16 (Continued)

Composition of compound	Melting point, °C	Type of crystal structure	Forbidden-band width ΔE, eV	Composition of compound	Melting point, °C	Type of crystal structure	Forbidden-band width ΔE, eV
$ZnSiAs_2$	—	Cub. (chalcopyrite)	2.10	$CuSbSe_2$	460	Rhomb.	—
$ZnGeP_2$	—	Same	2.20	$CuAsS_2$	625	—	—
$CdGeP_2$	—	»	1.80	$CuAsSe_2$	415	—	—
$ZnGeAs_2$	—	»	>0.60	$AgSbSe_2$	611	Cub.	—
$ZnSnAs_2$	—	»	—	$AgSbTe_2$	555	Cub.	—
$CdSnAs_2$	—	»	—	$CuSbSe_4$	425	Куб.	—
$CdSiP_2$	—	»	—	Ag_3AsS_3	480	—	—
$CuSbS_2$	535	Rhomb.	—	$CuSbSe_2$	472	—	—
$CdZnSb_2$	—	Hex.	—	$AgBiSe$	792	—	—
$CdZnSb_3$	—	—	—	α-β	298	—	—

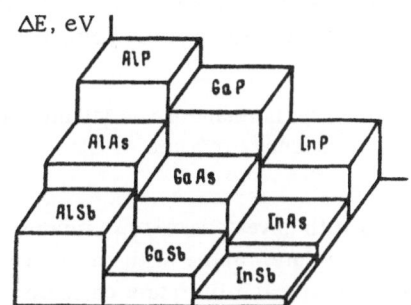

Fig. 15. Variation of forbidden-band width ΔE in the following three series of compounds: AlP—AlAs—AlSb; GaP—GaAs—GaSb; InP—InAs—InSb.

The compositions and some properties of semiconductor compounds formed by combining elements of different groups, taken from [105, 109], are given in Table 16. Data are grouped here on compounds of the types $A_{II}B_{IV}$ (where A is Mg or Ca), $A_{II}B_{VI}$ (where A is Zn or Cd), and $A_{III}B_V$ (where A is Al, Ga, In, or Tl), as well as various types of sulfides, selenides, tellurides, arsenides, antimonides, and bismuthides. The table also includes a group of ternary semiconductor compounds in which A is one of the elements Cu, Ag, Zn, Cd, In, etc.

Questions regarding certain conditions of formation of ternary and more complex compounds with semiconductor properties are considered in [65, 110]. The compositions of some compounds studied in [110] are also given in Table 16.

In all compounds of the groups considered above, one of the elements is a semiconductor. Many of these compounds are refractory and have isomorphous structures. Their forbidden-band width varies from a few hundredths of an electron volt to 3-4 eV.

The presence of isomorphous structures (mainly of the type of zinc blende ZnS), different degrees of metallic, covalent, and ionic bonding, and a variety of forbidden-band widths and interatomic distances in semiconductor compounds complicate the character of interactions between different groups of semiconductor compounds and make it difficult to determine the rules governing them.

Based on the data of Table 16, one may draw the important conclusion that as the electronegativity difference between elements A and B decreases, ΔE decreases and the proportion of metallic bonding in the compounds increases; as this difference increases, on the other hand, ionic bonding tends to predominate. This is confirmed by the data of Fig. 15 [104], where successive (stepwise) decreases in ΔE occur in the three series of compounds AlP—AlAs—AlSb, GaP—GaAs—GaSb, and InP—InAs—InSb. ΔE should be higher for boron compounds than for aluminum ones and lower for thallium compounds than for indium ones.

An article entitled "Intermetallic Compounds" appeared in an annual review on investigations in the field of physical chemistry in 1964 [268]. In this paper the author, mainly from results of foreign research, reviewed scientific work on intermetallic compounds (compounds of metals with metals). In the introduction to this article, attention was focused on the imperfection and confusion in terminology in the question as to what the term intermetallic compound means. Without making any recommendations regarding classification of metal compounds, but limiting himself to a discussion of intermetallide formation, the author excluded from consideration compounds of elements in Groups III-V and others, which are semiconductor compounds. As a whole, this review deals only with compounds of constant and variable compositions, formed by transition metals and lanthanides (with unfilled d and f electron shells) with other metals, i.e., purely intermetallic compounds.

In this review the author, considering the main factors determining the formation of such compounds and attaching much importance to the latest investigations in this field, devoted considerable attention to the electron theory of metals, magnetic and electrical properties, and investigations of compounds, using the Mössbauer effect. Among metal compounds, the author, based on the latest literature data, described certain chromium and manganese aluminides, questions of the distribution of valence electrons in the structures of such compounds, and, in connection with this, changes in magnetic and other properties. Compounds of iron with the Group IVB elements silicon, germanium, and tin — the so-called iron silicides, germanides, and stannides — were treated adequately. This group of compounds is especially interesting in the study of recoilless γ-ray absorption and emission, which are connected with the Mössbauer effect. Based on these studies, important conclusions were drawn regarding the nature of chemical bonding in iron silicides, germanides, and stannides; it proved to be mainly covalent. Investigations of chemical bonding in the metallides Mg_2Sn, $FeSn_2$, $PtSn_2$, and $InSn_2$, using the Mössbauer effect, were conducted also by Soviet authors [269-271]; based on these studies, which are not included in [268], the latter authors calculated the character of distribution of atoms in the structures of the compounds studied and established the nature of interatomic bonding in them.

Of the group of lanthanide compounds, the review deals with those formed by lanthanum and yttrium and manganese, iron, cobalt, and nickel, having various types of crystal structures. Recent literature on all these compounds is cited.

The review also contains information on metallides of composition AB_2 with Laves phase-type structures and deals with certain data on binary systems of metallides with this type of structure. Some of these metallide systems were not mentioned in the present monograph; hence we feel obliged to cite the literature on these topics. Thus, for instance, the author points out the studied systems $TiFe_2-ZrFe_2$ and $ZrFe_2-ZrCo_2$ [272], UFe_2-ZrFe_2 [273], and a number of alloys in binary metallide systems, having the following variable compositions: $A(B_{1-x}Si_x)_2$ [274] and $Zr(V_{1-x}Co_x)_2$ [275], where the components A and B in [274] are transition metals belonging to the first and second long periods.

The authors of [274] calculated the atomic radius of silicon in the studied compounds and showed that it varies from 1.16 to 1.21 A, i.e., it is much less than the value 1.34 A, calculated for pure silicon having the coordination number 12. Such decrease in the atomic radius of silicon is due to electron redistribution and change in the character of chemical bonding in these compounds. Further in this review the results of investigating many compounds — Nb_3Sn, Nb_3Al, V_3Ga, V_3Si, etc. — having superconductor properties are given, and questions regarding the superconductivity of intermetallides are discussed in connection with the concentration of valence electrons in them. The review also cites new papers describing the discovery and study of the following compounds of technetium with aluminum, molybdenum, tungsten, and rhenium: $TcAl_2$, $TcAl_4$, MoTc, TcW, and TcRe; the last three compounds are σ-phases with the structure $D8_V$.

Some questions of metal-compound formation are considered in [276] from the viewpoint of the disposition of components in the periodic system.

CHAPTER III

REACTIONS BETWEEN METALLIDES

Methods of Investigation

The materials considered above on metallide formation show that metallides are chemical individuals in many cases. The total number of such compounds is more than 4000, and it increases steadily as more and more new metallides are discovered in hitherto unstudied and little-studied metal—metal or metal—nonmetal systems (including here metal—gas systems). As shown above, daltonides are represented in the equilibrium diagrams by special singular points, whereas berthollides have a broad region of existence and are phases of variable composition. Proceeding from the stoichiometric compositions of daltonides, one should regard them as components in equilibrium systems. With a certain degree of arbitrariness, one may regard berthollides — phases of variable composition — as components.

In the field of metal chemistry an independent subdivision may be distinguished: metallide chemistry, dealing with chemical reactions between metallides.

Inasmuch as interactions of metals and equilibrium diagrams of metallic systems based on approximately 80 metallic elements have been successfully investigated, independent study of the interactions of more than 4000 metallides is quite natural.

Many compounds have individual, often complex, structures with different types of chemical bonding and special physical properties. All this makes questions of reactions between metallides particularly important.

Thus, from the viewpoint of their interactions, the chemistry of metallides should become the most extensive field of metal chemistry. It is more rich and varied in objects of investigation than the field of equilibria in simple metallic systems and much broader than the chemistry of halides or oxides, where ionic bonding alone prevails. Until recent times, experimental investigations in the field of reactions between metallides were conducted only in connection with the study of equilibrium diagrams of ternary and more complex metallic systems. Independent studies along this line were begun only recently. These are separate, uncoordinated efforts dealing with various groups of compounds; the number of metallide systems studied so far is limited.

Interest in reactions between various binary, ternary, and more complex metallides arose when the equilibrium diagrams of ternary and more complex metallic systems were first investigated. This was especially important for cases where a large number of binary and ternary metallic compounds are formed in the corresponding binary and ternary systems. Reactions between metallides in ternary systems are considered on the basis of phase-equilibrium theory by the triangulation method proposed by N. S. Kurnakov, G. G. Urazov, et al. [14-19]. This method is based on the fact that the formation of any binary or ternary compound (congruently melting) enables one to divide the entire ternary system into a number of partial ternary systems. Both the initial components of the ternary system as a whole and binary and ternary metallides may be components of such systems. The total system, from which the partial systems are obtained by triangulation, is called primary, whereas the partial systems are called secondary. Cases of triangulation of ternary systems on the basis of binary and ternary compound formation can be shown in many instances.

In the simple case where a single binary compound is formed in a ternary system, a so-called quasi-binary section can be taken through the single compound AB and the component C. This section divides the concentration triangle into two independent triangles A—AB—C and AB—B—C. They constitute two secondary ternary systems containing the same binary metallide.

In the case of a ternary system containing two binary compounds AB and AC (Fig. 16a.), several methods of triangulation may be used, one of which is isolation of the quasi-binary section passing through the two compounds AB and AC.

39

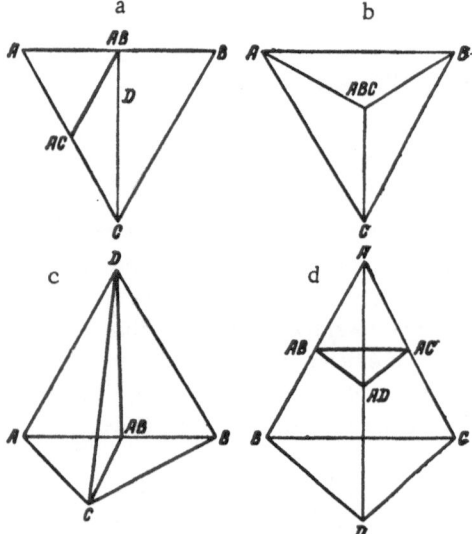

Fig. 16. Method of dividing ternary and quaternary systems containing compounds into secondary, binary and ternary systems: a) triangulation of ternary system in presence of single binary compound; b) same, in presence of single ternary compound; c) division of system containing single binary compound into tetrahedra; d) isolation of ternary system in presence of three compounds in quaternary system.

When ternary systems containing two such compounds are investigated, there are further possibilities of triangulation by passing quasi-binary sections through the compounds and metals. Of the possible cases shown in Fig. 16a, either of the systems AB—C and AC—B might be taken as such a section. One usually determines which system corresponds to reality by investigating experimentally the structure of the ternary alloy located at the intersection D of the two diagonals AB—C and AC—B. If the alloy of this composition consists of structures AB and C, the real equilibrium system is the quasi-binary system AB—C; if the structures AC and B are found in the alloy, the system AC—B is the real one.

Besides the above, purely empirical method of determining the quasi-binary character of sections by preparing an alloy and studying the structure of one composition of the ternary alloy (point D in Fig. 16a), this problem can be solved theoretically, by calculation. In our opinion, the latter should be based on thermochemical data and the relative stability of these two compounds having one element in common. Based on data on the heats of formation of the compounds AC and AB and their relative stability, as well as the following two displacement-reaction schemes:

$$AB + C = AC + B \qquad (I)$$
$$AC + B = AB + C \qquad (II)$$

one can answer the question as to which of these two quasi-binary sections is real and which is fictitious.

The above method enables one to triangulate, i.e., to isolate secondary ternary systems when two or more compounds are formed in each corresponding binary system.

The formation of a ternary metallide in a ternary system also enables one to triangulate the primary ternary system. In the simplest case, where only one ternary compound (ABC) is formed in the system and there are no compounds in the constituent binary systems, triangulation may be carried out as shown in Fig. 16b. In this case the diagonals ABC—A, ABC—B, and ABC—C are drawn through the composition point of the compound ABC and the primary ternary system then divided into three secondary ternary systems: A—ABC—B, A—ABC—C, and B—ABC—C. They are all independent systems each containing the ternary metallide. More complex triangulation methods, used in cases where several ternary compounds are formed and in the presence of binary compounds in ternary systems, are not considered here. They are well described in [14-19].

In studying reactions and equilibria between metallides in ternary systems, one is interested primarily in quasi-binary sections passsing through binary metallic compounds, and sections based on metallides whose components are pure metals.

The above factors, which determine the possibilities of reactions between metallides and formation of metallide solid solutions based on binary compounds, apply equally to the study of solid solutions based on ternary metallic compounds. In this case one must take more complex interactions into account, since atoms of more than three different kinds enter into the reaction. These interactions occur between components of primary ternary systems. Some examples of such complex systems will be considered below.

The main principles of classification of equilibria between metallides in binary and ternary systems can be extended to quaternary and more complex, many-component systems. Naturally, the number of possible binary, ternary, and more complex compounds increases with the number of components in the primary system; this complicates considerably the equilibrium systems under study.

For clarity in studying equilibria between metallides in a quaternary system, we shall consider the very simple case of formation of only one compound in one of the six binary systems making up this quaternary system. Such a case of a quaternary system A$-$B$-$C$-$D, where the compound AB (congruently melting) is formed in one of the systems A$-$B, is shown in Fig.16c. Taking this compound as a component, one can divide the primary tetrahedron ABCD into two secondary tetrahedra by passing a plane through the figurative point of compound AB and the edge CD connecting the vertices of components C and D, which do not occur in this compound. Such division of a quaternary system containing one binary compound into two parts with two tetrahedra ACDAB and BCDAB may be called "tetrahedration" in analogy with triangulation of a ternary system, considered above.

When several binary or ternary compounds are formed, the main rules of tetrahedration remain the same, but the phase diagrams of such systems become more complex.

To study reactions between metallides in quaternary and more complex systems, one can isolate by tetrahedration partial equilibrium diagrams of quasi-binary and quasi-ternary systems consisting of metallides. Such a case is shown in Fig. 16d, where three binary compounds of the same stoichiometric composition, having one common component, are formed in the quaternary system A$-$B$-$C$-$D. In this system, then, one may consider three quasi-binary systems made up of binary compounds: AB$-$AC, AC$-$AD, and AD$-$AB, and one quasi-ternary system of binary compounds: AB$-$AC$-$AD. Triangulation and tetrahedration are widely used in the physicochemical analysis and metal chemistry of many-component systems.

Main Factors Determining the Interaction of Metallides

Depending on the physicochemical nature and crystal structure of metallides, their interaction in binary systems, as in the case of simple metallic systems, may give: a) continuous solid solutions; b) limited solid solutions; c) mechanical mixtures without solid solutions; d) ternary metallides.

All these types of metallide interaction may be extended considerably by taking into account the interaction of metallides with one of the components of a ternary or more complex system. These cases are discussed in several books and papers [14, 15, 111-114]. In studying equilibria in metallide$-$metal (nonmetal) systems, cases of the formation of solid solutions, the absence of the latter, and the appearance of ternary compounds merit equal attention. However, these cases are not considered here.

The given types of interaction of metallic compounds correspond to the main types of equilibrium diagrams of metallide systems which are quasi-binary sections in ternary and more complex systems.

Three schematic equilibrium diagrams are given in Fig. 17 for the cases of formation between metallides of: continuous solid solutions (Fig. 17a), limited solutions (Fig. 17b), and simple eutectic mixtures without solid solutions (Fig. 17c). The formation of a ternary compound by two binary compounds introduces nothing essentially new, and it can be represented by these three diagrams. In contrast to solid solutions of metals, we proposed the name metallide solid solutions for solid solutions based on metallic compounds [40, 111-113]. What physicochemical factors determine the formation of metallide solid solutions (continuous and limited) in some systems and mechanical mixtures in others? One may draw the general conclusion that the main determining factors of solid solution formation between metallides are the positions of the elements in the periodic system, the electronic structure of the component atoms in the metallides, and the crystal structure of the compounds.

Continuous solid solutions of metallides appear under certain conditions in systems consisting of compounds formed by transformation from solid solutions (Kurnakov compounds) or on crystallization of melts. Among the latter they are formed by the interaction of berthollides, bertholide-daltonides, and daltonides.

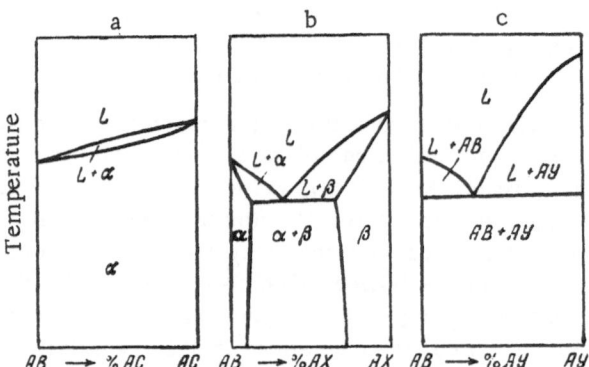

Fig. 17. Diagrams of binary metallide systems: a) continuous solid solutions; b) limited solid solutions; c) systems without solid solutions.

All three types of metallides form continuous solutions. These solutions are especially numerous among systems consisting of compounds such as carbides, borides, silicides, and also superconductor and semiconductor compounds.

The main factors determining the conditions of formation of continuous solid solutions are considered by us in [111-113].

By defining more precisely the theses set forth in these papers, one can arrive at the following conditions of formation of continuous metallide solid solutions:

1) The compounds must have the same crystal structure with nearly identical lattice parameters;
2) they must have the same type of chemical bonding;
3) they should contain elements with nearly identical atomic radii or one element in common;
4) they should have the same atomic ratio.

Of these four conditions, the first two are obligatory, whereas the third and fourth favor the formation of such solutions.

Thus in the system shown in Fig. 17a, compounds AB and AC must be isomorphous and contain atoms of elements which, with regard to their metallochemical properties, can be exchanged in the crystal structure of the compounds or can replace interstitial atoms with small radii.

It will be noted also that the character of metallide interactions and the corresponding equilibrium diagrams may be affected by polymorphic modifications of various compounds, the formation of ternary compounds, and other factors. It follows from these theses that compounds with different types of crystal structure or with polymorphic modifications, all other conditions being equal, cannot give continuous solid solutions owing to the impossibility of continuous transition from one structure type to another. As regards the type of chemical bonding in these compounds, it may be said that compounds with different types of bonding, e.g., metallic and covalent or metallic and partially or fully ionic, cannot form continuous mutual solid solutions. These theses will be confirmed by further description of experimental data.

The third condition for the formation of continuous metallide solid solutions is based on the fact that the two interacting compounds must contain atoms of analogous elements belonging to the same or nearby groups in the periodic system (e.g., Group VIII metals Fe, Co, Ni; Group V metals V, Nb, Ta, etc.), whose radii do not differ by more than 15%. When these conditions are met, continuous solid solutions are formed in such systems.

It is important for the continuity of metallide solid solutions to have atoms of analogous elements in the compounds; in some cases, indeed, one must have atoms of the same element. This will facilitate exchange of atoms of the same kind in the structure as the compositions of compounds in the metallide systems change.

A number of systems are known in which the presence of atoms of the same kind in different compounds (e.g., aluminides and borides of transition metals) favors the formation of continuous solid solutions by these compounds, other conditions being equal. Thus the third condition for the formation of continuous metallide solid solutions requires that the system contain either atoms of nearly identical structure or two compounds with one kind of atom in common. In a number of cases there are exceptions to these conditions, which will be discussed below.

The last condition — the presence of compounds having the same atomic ratio — follows from the fact that, as a rule, compounds with different atomic ratios have different types of crystal structure and chemical bonding; this hinders the formation of continuous solid solutions. When these compounds have the same types of structure and chemical bonding, they can form continuous solid solutions, as will be shown below in several examples of metallide systems.

Our theses on the formation of continuous metallide solid solutions were confirmed for the most part in later papers by many authors [65, 114-130].

In [65, 115, 116] the formation of continuous metallide solid solutions was first considered from the chemical viewpoint in the case of semiconductor compounds with covalent bonding. The results of these studies are generalized in a monograph [105].

Substitution of atoms and formation of continuous solid solutions between semiconductor compounds of elements in different subgroups, denoted by the general formulas $A_{II}B_{VI}$, $A_{III}B_V$, and $A_2^{III}B_3^{IV}$, were studied. A continuous series of solid solutions is formed by a pair of compounds of identical groups, having the same number of inner electrons, which the authors called isoelectronic compounds. In an investigation [115] of isomorphism between compounds with covalent bonding, the possibility of solid solutions in the systems cadmium telluride — zinc telluride (CdTe — ZnTe), cadmium telluride — mercuric telluride (CdTe — HgTe), InSb — GaSb, Ga_2Te_3 — In_2Te_3, and some others [105-106] was proved and their formation studied experimentally. These compounds with less pronounced covalent bonding form continuous substitutional solid solutions with relative ease. Compounds of the type of indium antimonide and arsenide (InSb and InAs), etc., whose bonding is more strongly covalent, tend to give continuous solid solutions only when they are annealed for a long time and their structures homogenized [105].

One may conclude from these studies that in metallic compounds with covalent bonding, the conditions for the formation of continuous solid solutions are more critical than in those with purely metallic bonding. As a rule, they do not form solid solutions with their components. ·

Developing these ideas, one may presume that compounds with different degrees of metallic and covalent or metallic and ionic bonding, i.e., compounds with different types of chemical bonding, do not form continuous solid solutions. These conclusions are of value for predicting the character of interaction between semiconductor compounds. In [115, 116] it is shown also that continuous solid solutions can be formed under certain conditions by isomorphous but heterovalent compounds of the types $A_2^{III}B_3^{VI}$ and $A^{III}B^{VI}$ (system Ga_2Se_3 — ZnSe), i.e., by compounds whose constituent elements have different valences.

Certain questions regarding the formation of continuous solid solutions by carbides, borides, silicides, and nitrides of transition metals are discussed in [118]. The author of this article, adopting for the most part our theses regarding the formation of continuous solid solutions by metallides [111], holds that the following conditions can be extended to the case of interstitial phases: 1) The compounds must have the same type of structure; 2) the constituent metals of the compounds must form continuous mutual solid solutions; 3) the compounds must have the same type of chemical bonding.

Literature materials on interactions of transition metal diborides are considered from this viewpoint in [119], and the conditions required for the formation of continuous solid solutions of interstitial phases are confirmed for the most part.

In [120] it was proved experimentally that continuous solid solutions can form in the system TiB_2-CrB_2 but do not occur in the system ZrB_2-CrB_2. The results of these investigations are in accord with the above rules of formation of continuous metallide solid solutions.

In an investigation of transition metal borides [119], structural characteristics are given for many borides with different atomic ratios, having hexagonal lattices, and the character of bonding between boron and metal atoms in such compounds is considered.

Systems of these borides contain metal atoms with unfilled d-electron shells and boron atoms. When the above conditions are met, according to [119], continuous solid solutions are formed in such systems only if the difference in atomic radius of the metals does not exceed 15%. If the difference is larger, only limited solid solutions are formed. Analogous conclusions were reached earlier in [131] in an investigation of the interrelations of phases in systems of titanium, zirconium, vanadium, and niobium nitrides and carbides.

Certain conditions determining the possibility of formation of limited metallide solid solutions are considered in [123]. The authors discuss cases of the formation of solid solutions based on metallides with different crystal structures. The limiting effect of a large difference in valence of the substituent atoms on the regions of solid solutions receives attention. It can be asserted that limited solid solutions are formed when the compounds are nonisomorphous and have different types of chemical bonding. The formation of metallide solid solutions with limited solubility should be expected when the constituent elements of the compounds differ in metallochemical properties and atomic radius (by more than 15%).

From the viewpoint of the conditions of formation of limited solid solutions, the following hypotheses regarding the formation of limited solid solutions based on the compound Ni_3Al, advanced in [132], are of interest. There it is noted, firstly, that those elements whose atomic radii are close to those of nickel or aluminum are the most soluble, these being Co, Fe, Cu, and Mn; secondly, that Co and Cu, whose electronic structures are close to that of nickel, replace nickel atoms in Ni_3Al; Si and Ti replace aluminum atoms, whereas Fe, Cr, and Mo replace atoms of both metals—Ni and Al.

The solubility of these elements in Ni_3Al depends on their position in various groups of the periodic system. When the above conditions for interaction of metallides are met, the equilibrium between them will be represented by a diagram (Fig. 17b) in which the compounds form limited solid solutions $AB-A_x$.

In the extreme case, where the differences in character of chemical bonding, size, and chemical properties of the constituent atoms in the compounds reach maximum values, conditions for the formation even of limited metallide solid solutions do not exist. In such cases metallide systems are represented, as shown in Fig. 17c ($Ab-A_y$), without any boundaries of limited component solubility.

Classification of Ternary Phases Based on Metallides

The question of classifying ternary phases based on metallic compounds received little attention until recent times, although they were investigated often. Some rules of formation of continuous ternary solid solutions based on binary metallic compounds were established in [40, 41, 111-114]. The chemical nature of a number of ternary berthollide and daltonide phases was considered in [113]; questions of the classification of some ternary phases from the viewpoint of similarity in electron concentration are discussed in [134, 135].

A more complete consideration of the physicochemical nature of ternary metallic phases, as well as a classification of them, based on comparison of the extent and character of disposition of regions showing homogeneity of ternary phases on the equilibrium diagrams, is given in [136, 137]. A development of this work was a review of the crystal structures of ternary metallic compounds with a classification of the latter according to crystallochemical features [41].

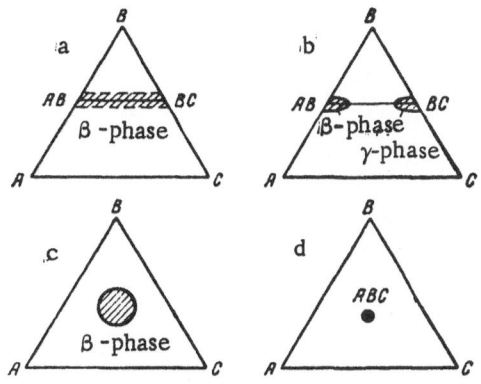

Fig. 18. Ternary metallic phases based on metallides: a) continuous solid solutions (β-phase); b) limited solid solutions (β- and γ-phases); c) ternary phase based on berthollide; d) ternary compound ABC—daltonide, without solid solutions.

In [137] the author, discussing the classification of ternary phases, proceeds from more general principles of their formation both from metals and from binary and ternary metallides. In this case he begins with Kurnakov's ideas regarding daltonides and berthollides [14, 15]. From this viewpoint these phases are divided into three groups:

1) ternary metallic phases of variable composition—solid solutions based on metals;

2) ternary solid solutions based on binary metallic compounds—metallide solid solutions;

3) ternary metallic compounds of variable or constant composition (berthollides and daltonides) and phases based on them.

Of these three groups of ternary metallic phases, the last two, which characterize compounds and phases formed from binary and ternary metallides, are related to the subject of this book. Of the total of eight types of ternary phases given in [136], therefore, phases based on metallic compounds include four main types. These types are shown in Fig. 18.

Ternary systems, in which two compounds of compositions AB and BC are formed in the constituent binary systems, include the first type (Fig. 18a). These compounds, being similar in metallochemical properties, may form continuous mutual solid solutions. Such solutions also may be formed by a binary compound and a pure-metal component of the system.

When the difference in the properties enumerated above is larger and the two compounds have different types of crystal structure, only limited solid solutions are formed, as is shown in Fig. 18b. Two ternary phases (γ and β), based on two compounds whose region of existence in the ternary system is limited, are formed in this case.

The third type (Fig. 18c) corresponds to phases with an extensive region of existence in the ternary system, which are based on a ternary compound of variable composition. Such phases may be ternary berthollides, which often are encountered in the corresponding ternary systems of aluminum, magnesium, etc. Examples are the ternary phase T(MgCuZn) in the system Mg—Zn—Cu, the phase U(AlMgCu) in the ternary system Al—Mg—Cu, etc.

Finally, the fourth type (Fig. 18d) of ternary phases comprises ternary metallic compounds of constant composition — daltonides, which do not form solid solutions with their components. In this case, as is evident from the figure, the position of this compound in the corresponding ternary system is characterized by a point. As will be shown below, these types of ternary phases based on binary and ternary metallic compounds have been confirmed experimentally in a number of systems.

The main principles of classification of phases in ternary metallic systems may be extended to quaternary and more complex systems. In them, too, secondary binary and ternary systems consisting of metallic compounds can be isolated by triangulation and tetrahedration, and ternary and more complex metallide phases found. These questions are considered in part in a textbook on the fundamentals of physicochemical analysis [15] and in other books [16-19].

CHAPTER IV

BINARY METALLIDE SYSTEMS WITH CONTINUOUS SOLID SOLUTIONS

The study of equilibria in metallide systems began in connection with the general development of investigations of equilibrium diagrams for ternary and many-component metallic systems. Today there are literature materials on a large number of binary, ternary, and more complex systems composed of metallides. Data on these systems are given briefly below.

Solid Solutions of Kurnakov Compounds

Let us consider first the literature on systems of Kurnakov compounds, which are formed by transformation of primary solid solutions, and then the literature on systems of metallides crystallized from the liquid phase.

One of the first groups is the ternary system Fe−Cr−V based on the metallides FeCr and FeV, which we studied [45]. As is well known [43], the compounds FeCr and FeV are formed in the binary system Fe−Cr and Fe−V from continuous α-solid solutions. According to both chemical and structural criteria, these two compounds satisfy the main conditions of formation of continuous mutual solid solutions. An experimental investigation of the quasi-binary section of the ternary system Fe−Cr−V, passing through the compositions FeCr and FeV, showed that ternary alloys of this section crystallize as continuous α-solid solutions [45]. On slow cooling or prolonged maintenance below the critical temperatures of formation of the compounds, the α-solid solutions are transformed into metallide solid solutions FeCr−FeV, which are called the σ-phase. In the section studied, this $\alpha \rightleftharpoons \sigma$ phase transition takes place along a continuous curve from the compound FeCr to FeV.

The equilibrium diagram for alloys of the section FeCr−FeV (Fig. 19) shows that the alloys crystallize as continuous ternary α-solid solutions with a body-centered cubic structure and are transformed to σ-solid solutions in a certain interval. The σ-phase has a tetragonal, β-U-type structure. The continuous character of the metallide solid solutions was confirmed by x-ray diffraction, as well as hardness and electrical-resistivity measurements. In this region of continuous σ-solid solutions (see the bottom of Fig. 19), the properties studied vary continuously through a relatively flat maximum. As was shown in [45], the ternary Fe-α solid solution and the σ-solid solution (compounds CrFe and VFe), which differ fundamentally in physicochemical nature and structure, are markedly different in properties. The hardness H_v of the σ-solid solution varies with composition from 600 to 900 kg/mm^2, whereas for α-solid solutions of iron (in the quenched state) in the same composition interval the hardness varies only from 270 to 420 kg/mm^2 (see Fig. 19, bottom).

Many cases can be cited in which Kurnakov compounds form continuous mutual solid solutions. One such case is the transition from FeNi$_3$ to CrNi$_3$ through a continuous series of solid solutions of these metallides, formed from primary solid solutions of the systems Fe−Ni and Ni−Cr [138].

In a study of the ternary system VNi$_3$−VCo$_3$−VPd$_3$ [139], it was shown that continuous solid solutions are formed in the systems of isostructural compounds VNi$_3$−VCo$_3$ and VNi$_3$−VPd$_3$. In these systems, transitions from ternary metal solid solutions to metallide ones also take place along continuous curves. In the third binary system studied by the authors, VCo$_3$−VPd$_3$, a break in solubility and a two-phase region were found. This is explained by the break in continuity of solid solutions in the binary system Pd−Co and the formation of additional intermediate phases in the latter.

The formation of continuous solid solutions of Kurnakov compounds may be presumed in many systems which have not yet been studied experimentally. One is just as likely to find Kurnakov compounds formed from continuous solid solutions as from limited ones.

47

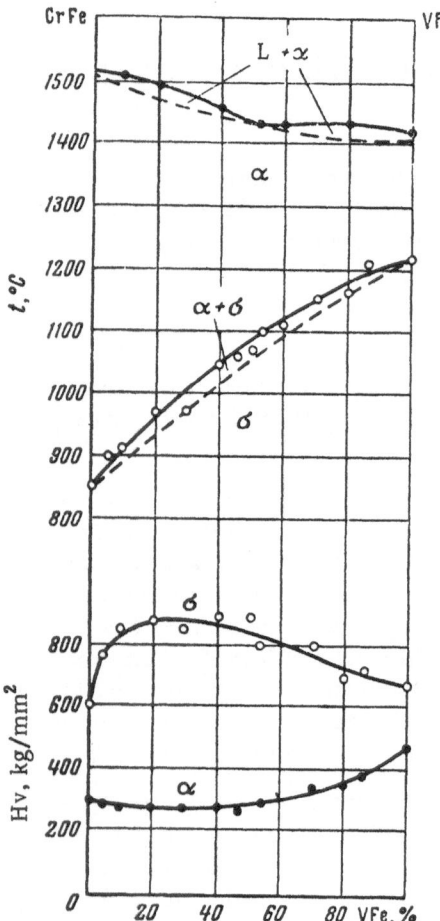

Fig. 19. Quasi-binary system of section FeCr—FeV: top) according to thermal analysis data; bottom) properties of annealed alloys in σ-phase state and quenched alloys in α-phase state.

In Table 17 are given systems composed of Kurnakov compounds, in which continuous solid solutions either have been found experimentally [138-139] or should be expected [111-113]. They are given for compounds of three structure types: face-centered cubic, tetragonal, and β-U (σ-phase).

Further experimental study of such systems should confirm the existence in them of continuous solid solutions whose properties vary characteristically with composition.

One interesting case, where a Kurnakov compound forms continuous solid solutions with a daltonide crystallizing from the liquid phase, is the quasi-binary system Ti$_3$Al—Ti$_3$Sn, which occurs in the ternary system Ti—Al—Sn with limited solid solutions in α-Ti [40].

In several literature sources [43, 46] it is stated that the compounds Ti$_6$Al, Ti$_3$Al, and Ti$_2$Al may be formed from α-solid solutions of Ti in the system Ti—Al. These statements are based on data obtained by studying various properties of this binary system. In [46], as was shown above (see Fig. 4), the compositions of the compounds Ti$_6$Al and Ti$_3$Al were found from inflection points on the Hall effect — composition diagram. Thus it may be presumed that one of these compounds — Ti$_3$Al — in the system Ti—Al is formed from α-solid solutions of titanium, having an ordered structure. This compound has a hexagonal structure and is isomorphous with Ti$_3$Sn.

The composition, structure, and properties of Ti$_3$Sn (45.24 wt.% Sn) were considered in [43]. This compound melts with a real maximum at 1663° and has a hexagonal close-packed structure.

The formation of a considerable region of ternary solid solutions based on α-titanium and Ti$_3$Sn, having a hexagonal structure, was established when the equilibrium diagram of the ternary system Ti—Al—Sn was investigated by microstructural and x-ray methods [140].

In [141] the phase diagram of the ternary system Ti—Al—Sn was studied, as well as the properties of alloys of this system, corresponding in composition to alloys in the quasi-binary section Ti$_3$Al—Ti$_3$Sn. The alloys were studied by thermal, microstructure, and x-ray analysis and electrical-resistivity and hardness measurements.

After analyzing the thermograms we plotted the liquidus and solidus lines of alloys in the section Ti$_3$Al—Ti$_3$Sn and noted solid-state transformation points for alloys having polymorphic or other types of transformation. The phase diagram plotted from thermal analysis data shows that the alloys crystallize as eutectics with mutually limited solid solutions, as shown in Fig. 20, top.

Most of the alloys studied are transformed in the solid state owing to the polymorphism of titanium and the formation of solid solutions of the two compounds Ti$_3$Al and Ti$_3$Sn. For the alloys of composition Ti$_3$Al the transition from the region α(δ) to the region α(δ) + β begins at a temperature of 1120°. On addition of Ti$_3$Sn to Ti$_3$Al this temperature decreases slightly at first from that of α ⇌ β-transformation to a minimum of 990° at 60% Ti$_3$Sn, and then rises to 1035° (see Fig. 20).

On thermograms of alloys in the section Ti$_3$Al—Ti$_3$Sn, close in composition to Ti$_3$Al, additional heat effects are manifested at temperatures below 960°. They are denoted in Fig. 20 by the dashed line. These effects

TABLE 17. Systems with Continuous Solid Solutions of Kurnakov Compounds

Structure type and system	Literature	Structure type and system	Literature
I. AuCu$_3$ (f. c. c.)		II. AuCu (tetrag.)	
AuCu$_3$ — PtCu$_3$	[111]	AuCu — PtCu	[111]
AuCu$_3$ — PdCu$_3$	Hypoth.	AuCu — PdCu	Hypoth.
PtCu$_3$ — PdCu$_3$	»	PdCu — PtCu	»
MnNi$_3$ — FeNi$_3$	»	FePd — FePt	»
MnNi$_3$ — CrNi$_3$	»	III. β -U (σ -phase)	
FeNi$_3$ — CrNi$_3$	[138]	CrFe — VFe	[45]
VNi$_3$ — VCo$_3$	[139]	CrFe — VCo	[111]
VPd$_3$ — VNi$_3$	[139]	CrFe — CrCo	Hypoth.
VNi$_3$ — FeNi$_3$	[111]	CrCo — MoCo	»
VNi$_3$ — CrNi$_3$	Hypoth.		

apparently are due to the formation of solid solutions of Ti$_3$Al andTi$_3$Sn from α(δ)-solid solutions of titanium. The temperature of formation of Ti$_3$Al in the binary system Ti—Al, according to thermal analysis data, is 960°. This temperature decreases gradually with increasing Ti$_3$Sn concentration, as the dashed line shows.

The continuous series of solid solutions of annealed alloys in the section Ti$_3$Al—Ti$_3$Sn is confirmed by the curves of hardness and electrical resistivity versus composition, shown in Fig. 20. These two curves have slopes of the same sign and in the property—composition diagram pass through a flat maximum which is characteristic of systems containing continuous metallide solid solutions.

The continuous character of solid solutions of these metallides is shown also by the heat resistance—composition diagram for alloys of the system Ti$_3$Al—Ti$_3$Sn in the temperature interval 700-750°. Curves of time, required for the flexure indicator to travel 2, 3, 4, 5, and 7 mm at 700-750° under 20-25 kg/mm^2 stress, versus composition have flat maxima corresponding to the Ti$_3$Al alloy containing 50% Ti$_3$Sn (Fig. 21).

It is known from the literature [43] that the compound Ti$_3$Sn is isomorphous with Ti$_3$Al and Ti$_3$In. It may be presumed from this that solid solutions can exist in the metallide system Ti$_3$Al— Ti$_3$Sn and Ti$_3$In—Ti$_3$Sn. These questions require further investigation.

In connection with the above experimental materials on solid solutions of Kurnakov compounds, a paper [227] devoted to binary systems of such compounds with ordered structures, CoPt—CrPt and CoPt—MnPt, is of interest. The authors studied the formation of solid solutions with disordered (above the critical transformation point) and ordered structures in a number of alloys of these systems, as well as changes in the magnetic properties of such alloys with composition.

In an investigation of the first metallide system, CoPt—CrPt, mainly by x-ray methods, the authors, in ac-

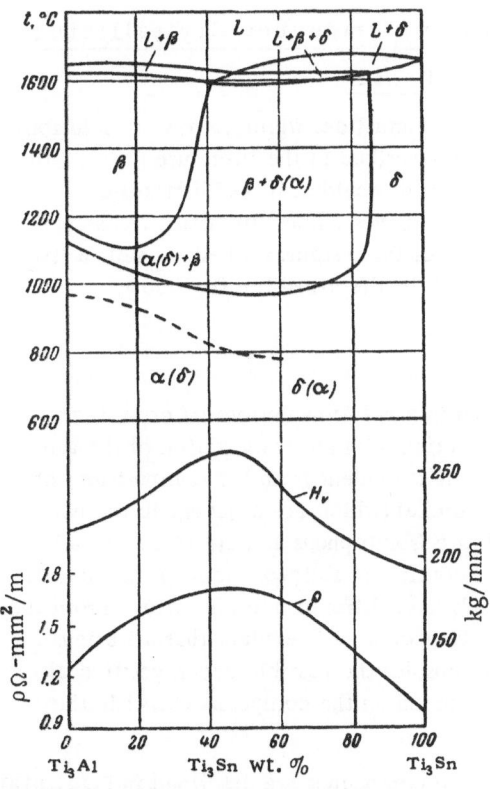

Fig. 20. Quasi-binary system of section Ti$_3$Al—Ti$_3$Sn: top) from thermal analysis data; bottom) properties of annealed alloys.

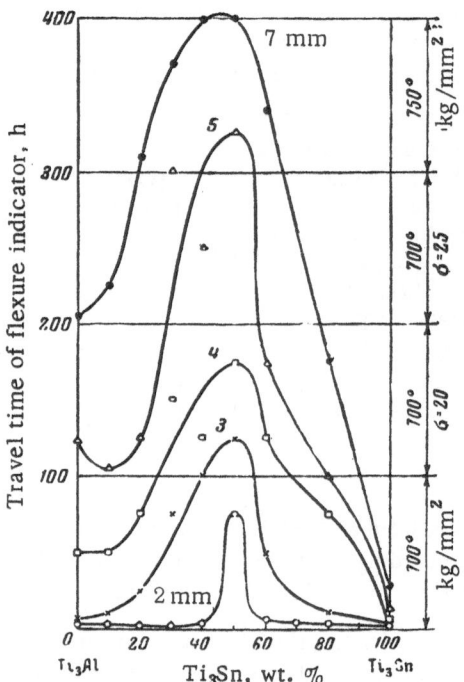

Fig. 21. Diagram of time required for flexure indicator to reach 2, 3, 4, 5 and 7 mm composition according to test data on heat resistance of alloys of system Ti$_3$Al—Ti$_3$Sn, obtained by centrifugal method.

cordance with the rule given above, established the formation of continuous solid solutions between these compounds. The lattice parameters and critical points of compound formation vary linearly with composition. Data on two types of ordering reactions in these compounds and their solid solutions are of interest. At relatively high temperatures (about 700°) these reactions are continuous, whereas at about 500° they are discontinuous. In the continuous reaction the system gradually goes over from the cubic structure of the disordered ternary solid solution Co—Pt—Cr to the ordered solid solution of the metallides CoPt and CrPt. In the discontinuous reaction the cubic structure persists for a long time together with the well-marked tetragonal structure of the ordered phase.

The second system, CoPt—MnPt, proved to be more complicated than the first owing to the complex structure of the compound MnPt at high temperatures and its conversion to another modification at 950°. As investigations by the authors showed, a break in solubility occurs in this system owing to nonisomorphism of the compounds, and solid solutions on the CoPt side are limited to a maximum concentration of about 15-16% Mn. Above this content two phases are formed, the second phase being a solid solution of CoPt in MnPt.

Solid Solutions of Compounds Crystallizing

From the Liquid Phase

Some systems of metallides which form solid solutions on crystallization are described in the literature [15-19, 40, 53-57, 111-113]. In this connection there are many data on berthollides, daltonides, mixed berthollide-daltonides, various classes of electron compounds, nickel arsenide phases, Laves phases, interstitial phases, compounds of the superconductor and semiconductor groups, etc. Some of the systems studied, as well as systems in which the presence of such solid solutions may be presumed, will be described briefly below.

Solid Solutions of Electron Compounds

One of the first metallide systems in which the formation of continuous solid solutions on crystallization was established was the system Cu$_3$Al—NiAl, studied in 1923 in connection with an investigation of the ternary system Cu—Ni—Al [142]. Both components are electron compounds. As is evident from the compositions of these compounds, having an aluminum atom in common, copper and nickel (which are adjacent in the periodic system) form continuous mutual solid solutions. The two compounds are isomorphous and crystallize in a body-centered cubic structure. All these conditions favor the formation of continuous solid solutions in the metallide system Cu$_3$Al—NiAl. This takes place even though the two compounds have different atomic ratios. From the fusibility diagram of this system, plotted in Fig. 22 from thermal analysis data, it is evident that all alloys crystallize as a continuous series of solid solutions. Here we do not consider the possible decomposition of these solid solutions in connection with decomposition of the β-phase based on the compound Cu$_3$Al in the solid state.

Systematic investigations in the field of solid solutions of electron compounds are described in [114, 143]. In an article on the mutual solubility of electron compounds [143] it is noted in the cases of silver alloys with cadmium and zinc that such compounds form continuous mutual solid solutions provided that these phases are isostructural and have the same electron concentration. All solid solutions of these compounds are substitutional

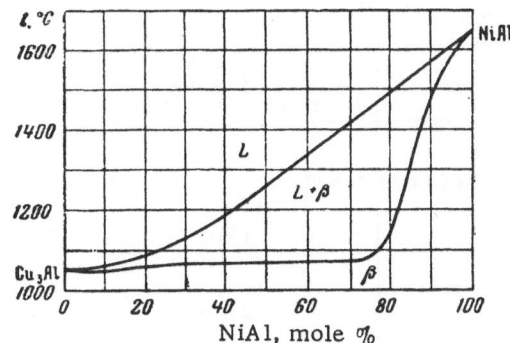

Fig. 22. Fusibility diagram of metallide system Cu$_3$Al$-$NiAl.

and have metallic bonding. Substitution of atoms in the structure of such compounds conforms to the principle of similarity of metallochemical properties, e.g., Cu and Ni, Fe and Ni, Fe and Co, Zn and Cd, Cd and Mg, Be and Al. In this case the difference in atomic radius must not exceed 17% for the cubic structure, or 11% for the hexagonal.

The results of an experimental study of interactions between phases in the ternary system Ag$-$Zn$-$Cd are published in [124, 143]. The phase diagram of this system is plotted from these data in Fig. 23, the regions of existence of metallide solid solutions being indicated. As is evident from the diagram, the β-electron compounds AgZn and AgCd form continuous solid solutions at temperatures above 500°. At 400° AgCd goes over from the body-centered β-phase to the hexagonal β'-phase, and hence there is a gap in the mutual solubility of these two compounds. The presence of continuous γ-solid solutions of isomorphous compounds in the system Ag$_5$Zn$_8$$-Ag_5Cd_8$ also was confirmed (see Fig. 23). However, study of alloys of the system AgZn$_3$$-$AgCd$_3$ showed that even when these compounds (hexagonal) are isomorphous, they do not form continuous mutual solid solutions (see Fig. 23).

In the author's opinion, this is due to the high content (75 mole %) of substituent atoms of Zn and Cd which do not form continuous mutual solid solutions. Moreover, it should be added that in these compounds of complex structure the interatomic distances could change; this would result in limited mutual solubility of their constituent elements.

Analogous cases of formation of continuous solid solutions were found in the ternary systems Ag$-$Cd$-$Mg, Ag$-$Cu$-$Zn, and Ag$-$Mg$-$Zn, where electron compounds also are formed [114]. The presence of the following continuous solid solutions in these systems at high temperatures was established: AgCd$-$AgMg, β-AgCd$-\beta$-AgZn, β-CuZn$-\beta$-AgZn, and AgMg$-$AgZn. The possible formation of such solutions in the system AuZn$-$CuZn is indicated in [24]. At low temperatures there are breaks in continuity owing to transformation of the compounds into other crystalline modifications and their decomposition into other phases. There are many such examples of the formation of continuous mutual solid solutions of electron compounds. Binary systems of β-, γ-, and ε-electron compounds (with different electron:atom ratios e/A), taken from [143], are given in Table 18, together with the kinds of atoms exchanged in solid solution formation and the relative differences in their size, in percent.

As is evident from Table 18, compounds having different atomic ratios form continuous solid solutions when they are isomorphous and their constituent elements have nearly identical metallochemical properties. Atoms of different kinds are exchanged in such compounds when the atoms either are similar in kind (Al and Zn), are fully analogous (Fe and Co, Fe and Ni, and Ni and Co), or have similar chemical properties (Be and Al, Mg and Zn, Zn and Cd, etc.). Thus the above

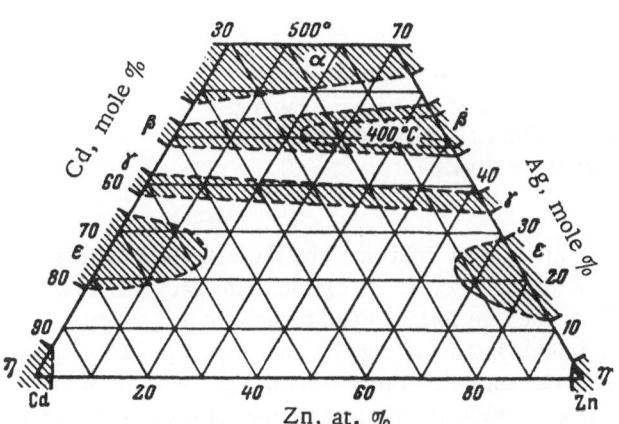

Fig. 23. Regions of metallide solid solutions in ternary system Ag$-$Zn$-$Cd.

TABLE 18. Continuous Mutual Solid Solutions of Electron Compounds [143]

Metallide system	Metals exchanged	Relative difference in size of atoms exchanged*, %
Cubic structure β-electron compounds with ratio e/A = 3/2		
AgMg — AgZn	Mg — Zn	14.4
Cu$_3$Al — CuZn	Zn — Al (partly Cu)	1.5
Cu$_3$Al — CuBe	Be — Al	17.0
Cu$_3$Al — Cu$_5$Sn	Al — Sn	6.4
Cu$_3$Al — NiAl	Cu — Ni (partly Al)	2.2
CuZn — NiZn	Cu — Ni	3.8
» — MnZn	Cu — Mn	6.5
NiAl — FeAl	Ni — Fe	1.6
» — MnAl	Ni — Mn	5.3
MnAl — CoAl	Mn — Co	3.8
FeAl — CoAl	Fe — Co	0.8
FeAl — NiAl	Fe — Ni	—
NiAl — CoAl	Ni — Co	—
AgZn — CuZn	Ag — Cu	11.1
γ-electron compounds with e/A = 21/13		
Cu$_5$Zn$_8$ — Cu$_9$Al$_4$	Zn — Al	1.5
» — Ag$_5$Zn$_8$	Cu — Al (partly Ag)	11.1
» — Ni$_5$Zn$_{21}$	Cu — Ni	2.4
Hexagonal structure ε-electron compounds with e/A = 7/4		
AgZn$_3$ — Ag$_5$Al$_3$	Zn — Al (partly Ag)	4.5
» — CuZn$_3$	Ag — Cu	11.1
» — MnZn$_7$	Ag — Mn (partly Zn)	4.9
CuZn$_3$ — MnZn$_7$	Cu — Mn	6.5

*The size factor was calculated by the formula $F = (r_A - r_{A'})/r_A \cdot 100$ or $F = (r_B - r_{B'})/r_B \cdot 100$. where r_A and r_B are the radii of the larger exchanging atoms. When two atoms are substituted simultaneously, the size factor F is calculated from the average of their radii.

conditions of formation of continuous metallide solid solutions are mainly confirmed by the examples of these compounds. However, there are some exceptions to these conditions. Such an exception is the fact that in a number of systems of electron compounds, continuous solid solutions are formed even though the exchanging atoms in these compounds do not form such solutions (e.g., Be and Al, Zn and Mg, Zn and Cd, etc.). This apparently is explained by the fact that all electron compounds have purely metallic bonding, and when atoms of different kinds but with similar metallochemical properties are substituted in the structure of such compounds, the sizes of the diverse atoms and the character of their arrangement change in the direction favoring their continuous substitution, at least at high temperatures. These questions require further study. Constancy of the ratio

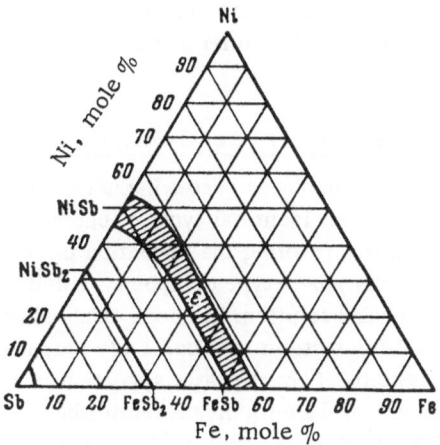

Fig. 24. Continuous solid solutions of ε - metallides NiSb—FeSb in ternary system Fe—Ni—Sb.

of the number of electrons to number of atoms in the case of mutually soluble electron compounds having fully isomorphous structures favors the formation of continuous solid solutions.

Solid Solutions of Nickel Arsenide-Type Compounds

The large class of nickel arsenide-type compounds considered above is encountered in many metallic systems. These compounds form continuous solid solutions when the conditions enumerated above are met. Continuous NiSb—FeSb solid solutions in the ternary system Fe—Ni—Sb are a classic example of the formation of phases of variable composition. This system was studied in 1941 and described in [24, 144].

The phase diagram of this system at room temperature is shown in Fig. 24. NiSb and FeSb both crystallize as nickel arsenide-type phases; they are isomorphous and, besides the common Sb atom, contain analogous metals (nickel and iron), which is one of the main conditions for continuity of metallide solid solutions. As is evident from Fig. 24, a continuous transition actually takes place in this ternary system— from NiSb to FeSb through the variable-composition phase ε, which N. V. Ageev called a Kurnakov phase. This case of formation of continuous solid solutions is interesting, in that NiSb is a typical daltonide, whereas FeSb is a berthollide. The system NiSb—FeSb is a striking example of the formation of continuous daltonide—berthollide solid solutions.

On this basis one may conclude that when the main conditions for the formation of continuous solid solutions are met, daltonides can form such solutions with berthollides.

Only a limited number of systems of metallides with the NiAs structure, forming continuous solid solutions, have been studied. They are described in [123, 124]. Data on continuous solubility in some of the systems studied, taken from the cited sources, are given in Table 19.

As is evident from Table 19, the metallide systems given in it (CrSb—MnSb, CrSb—FeSb, etc.) satisfy the main conditions for the formation of continuous metallide solid solutions. In the studied systems having a common atom (Sb) or analogous atoms (Sb and Bi), the other atoms also are analogous (Fe, Ni, Co) or close in metallochemical properties (Cr, Mn and Cr, Fe); this no doubt favors continuous substitution of atoms in the isomorphous structures of the compounds. The system NiSb—Ni_3Sn_2, which was studied in [145], is interesting. Of these two compounds NiSb, as is well known, is a daltonide and Ni_3Sn_2 a berthollide. It was proved by investigation that a continuous series of solid solutions is formed in this system. The above conditions of formation of continuous solid solutions do not apply to this system, since, firstly, these compounds do not have identical atomic ratios, and secondly, Sb and Sn do not form continuous mutual solid solutions.

The observed effect may be due to the fact that antimony and tin lie in the same (seventh) row in the fifth period, are adjacent in the periodic system, and have nearly identical electronegativities and atomic radii (within 1.9%). When the compounds NiSb and Ni_3Sn_2 are isomorphous, all this will

TABLE 19. Continuous Solid Solutions of Compounds of the Type of Nickel Arsenide Phases

Metallide system	Relative difference, %	
	in atomic radius	in length of period
CrSb—MnSb	6.6	2.2
CrSb—FeSb	1.6	0.7
MnSb—FeSb	8.0	2.9
CoSb—NiSb	0.8	0.8
NiSb—NiBi	6.5	3.0
NiSb—Ni_3Sn_2	1.9	4.1
$Fe_{1.27}Sn$—Ni_3Sn_2	2.0	—

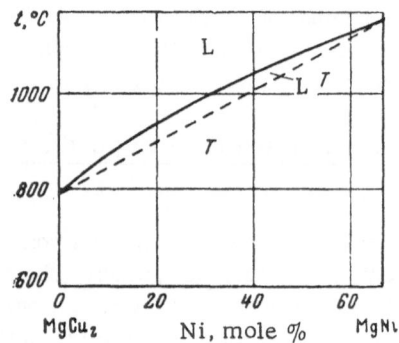

Fig. 25. Fusibility diagram of system $MgCu_2-MgN_2$.

favor the exchange of Sb and Sn atoms in the structures of these compounds. These exceptions to the general rules and the causes of the deviation will be discussed below.

Solid Solutions of Laves Phase-Type Compounds

As do other classes of compounds, Laves phases include some compounds with isomorphous structures, identical types of chemical bonding (metallic), and atomic ratios favoring the formation of continuous solid solutions. Of the Laves phase-type compounds listed above (Table 6), the presence of solid solutions has been established as yet in only a few of the metallide systems studied. Only recently investigations of a number of systems of these phases appeared in which the formation of continuous solid solutions was established. One of these systems, which were studied in [125], is the $MgCu_2-MgNi_2$. According to structural analysis data, compounds of this system have either of two types of structure: $MgCu_2$ has a cubic structure and $MgNi_2$ a hexagonal. Despite the difference in structure type of these compounds, the system $MgCu_2-MgNi_2$ crystallizes, according to thermal analysis data [125], as a continuous series of solid solutions (Fig. 25). As is evident from the compositions of these compounds, the conditions for this exist, structure type excepted. The break in continuity due to the difference in structure type found from structural type analysis data [146], requires further study.

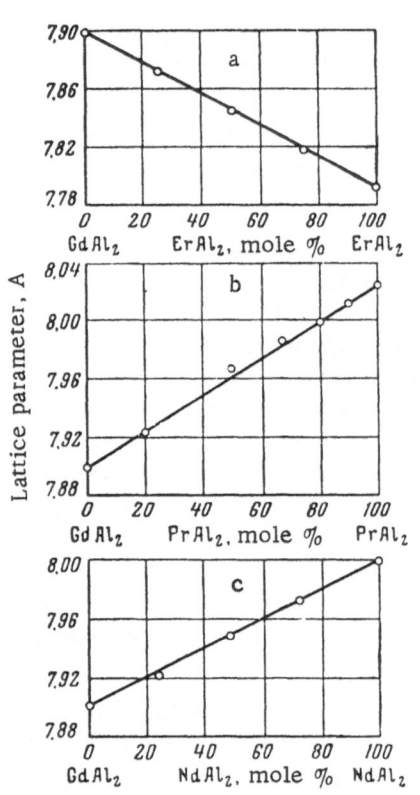

Fig. 26. Lattice parameters of systems at 25°: a) $GdAl_2-ErAl_2$; b) $GdAl_2-PrAl_2$; c) $GdAl_2-NdAl_2$.

Compounds of rare-earth metals with transition metals are an interesting example of the formation of continuous solid solutions between Laves phases. A paper devoted to these investigations was published recently under the title, Pseudobinary systems of Laves phases of rare-earth metals [147]. The authors, using x-ray structure analysis for the most part, established the presence of continuous solid solutions in many of the metallide systems studied. In composite Table 20 are given metallide systems, 12 of which were studied in [147]. They are systems of compounds AB_2, where A is any one of the metals dysprosium, holmium, gadolinium, terbium, etc., and B is Mn, Al, or one of the metallochemically similar metals Mn, Fe, and Ni. Along with structural isomorphism and identical types of chemical bonding, all other conditions for the formation of continuous solid solutions between the compounds are met. The continuity of these solutions is confirmed by the smoothness and linearity of the curve of lattice parameter versus composition, as shown in Fig. 26 for the systems $GdAl_2-ErAl_2$, $GdAl_2-PrAl_2$, and $GdAl_2-NdAl_2$.

Certain deviations of the lattice parameter—composition curve from linearity in some systems attest, as the authors correctly note, to electron interaction between constituent Al and Mn atoms and show a tendency toward the formation of a new phase.

All these compounds have a $MgCu_2$-type structure (cub. C15), i.e., they are isostructural. Several systems of this series form limited solid solutions, which will be discussed below.

One newly investigated metallide system with continuous solid solutions is the system $TiFe_2-TiCr_2$, which was studied in [148].

TABLE 20. Systems in which Laves Phase-Type Compounds, Having the General
Formula AB_2, Form Continuous Solid Solutions

System	Literature	System	Literature
Compounds of rare-earth metals			
MgCu$_2$ type (cub. C15)		MgCu$_2$ type (cub. C15)	
DyMn$_2$ — HoMn$_2$	[147]	GdAl$_2$—ErAl$_2$	[147]
HoMn$_2$ — HoFe$_2$	»	GdAl$_2$ — PrAl$_2$	»
DyMn$_2$ — DyFe$_2$	»	GdAl$_2$ — NdAl$_2$	»
HoMn$_2$ — HoAl$_2$	»	GdAl$_2$ — DyAl$_2$	»
TbMn$_2$ — TbAl$_2$	»	TbAl$_2$ — NdAl$_2$	»
DyMn$_2$ — DyAl$_2$	»	TbAl$_2$ — DyAl$_2$	»
Systems in which the formation of continuous solid solutions is to be expected			
MgCu$_2$ type		MgZn$_2$ type	
BaPd$_2$ — BaPt$_2$	Hypoth.	SrMg$_2$ — BaMg$_2$	Hypoth.
CaPd$_2$ — CaPt$_2$	»	NbFe$_2$ — TaFe$_2$	»
CeCo$_2$ — CeNi$_2$	»	TaFe$_2$ — WFe	»
HfCr$_2$ — HfMo$_2$	»	VBe$_2$ — CrBe$_2$	»
HfMo$_2$ — HfW$_2$	»	VBe$_2$ — MoBe$_2$	»
HfMo$_2$ — HfV$_2$	»	MoBe$_2$ — VBe$_2$	»
NaAg$_2$ — NaAu$_2$	»	MoBe$_2$ — WBe$_2$	»
TaCo$_2$ — NbCo$_2$	»	WBe$_2$ — ReBe$_2$	»
TaCr$_2$ — TiCr$_2$	»		
UCo$_2$ — UFe$_2$	»		
ZrCr$_2$ — TiCr$_2$	»		
ZrCr$_2$ — ZrFe$_2$	[153]		
(break in continuity on account of modification of ZrCr$_2$)			

Both TiFe$_2$ and TiCr$_2$ are Laves phases. TiFe$_2$ melts congruently at ~1500° and has a MgZn$_2$-type (hexagonal) structure whose lattice parameters are: a = 4.779 A, c = 7.760 A; c/a = 1.62. TiCr$_2$ is formed on decomposition of a continuous series of solutions based on β-titanium and chromium. Its temperature of formation is 1360°. According to [149], it exists in two polymorphic modifications: Above 1220° the MgZn$_2$-type (hexagonal) structure is stable, whereas below this temperature the MgCu$_2$-type (cubic) is. The lattice parameters of the high-temperature modification are: a = 4.921 A, c = 7.984 A; c/a = 1.62. As is evident, they are close to the lattice parameters of TiFe$_2$ (see above).

Thus since the titanium atom is common to both these compounds and the Fe and Cr atoms in them are metallochemically similar, the high-temperature modification of TiCr$_2$ is still isomorphous with TiFe$_2$.

Even in [150] it was concluded from x-ray measurements that TiFe$_2$ and TiCr$_2$ formed a continuous series of solid solutions with a MgZn$_2$-type structure at temperatures above 1100°. An investigation of the interaction of TiFe$_2$ and TiCr$_2$, conducted by thermal, microstructure, and x-ray analysis, proved unambiguously that they form continuous mutual solid solutions.

The equilibrium diagram of the system TiFe$_2$—TiCr$_2$ is shown in Fig. 27. The solidus line of this system falls smoothly from the melting point of TiCr$_2$ to those of alloys containing 60-65% Cr, which begin to melt at 1360-1380°, and then rises to the solidus point of an alloy in the binary system Ti—Cr. Castings of alloys containing 5-50% Cr have a microstructure characterizing crystallization of solid solutions based on TiFe$_2$ and TiCr$_2$. In the case of alloys containing 60 and 65% Cr, a small amount of a second constituent phase is formed

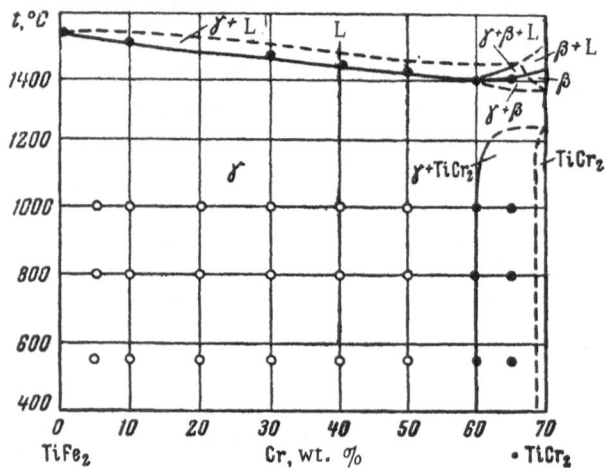

Fig. 27. Polythermic equilibrium diagram of system TiFe$_2$— TiCr$_2$.

at the boundaries of the primary crystals, which corresponds to residual β-solid solutions of Ti and Cr. It decomposes during annealing at 1000°, as a result of which the amount of it present is even smaller. X-ray analysis of alloys containing up to 50% Cr showed that the investigated alloys all have the same system of reflection lines of the MgZn$_2$-type structure. In the case of the alloy containing 65% Cr these lines are diffuse, and weak reflection lines appear corresponding to the low-temperature modification of TiCr$_2$, which has a MgCu$_2$-type structure. These lines are more intense after annealing at 550°.

According to the literature [149] and our data [148], the lattice parameters a and c for the alloys investigated increase steadily from the values for TiFe$_2$ to those for TiCr$_2$; this confirms the formation of a continuous series of solid solutions between the isomorphous compounds TiFe$_2$ and TiCr$_2$ at high temperatures. The solid solution based on these compounds is a Laves phase Ti(CrFe)$_2$ of variable composition, which we denote by the term γ-phase, where there is a common titanium atom, and iron and chromium atoms are distributed at random throughout the structure. The presence of the cubic form of TiCr$_2$ at low temperatures causes a break in continuity and gives rise to the two-phase region γ + TiCr$_2$ (cub.) (see Fig. 27).

Another interesting example of the formation of continuous solid solutions of Laves phases is the system TiCr$_2$—NbCr$_2$ [151, 152]. This system is interesting in that TiCr$_2$ and NbCr$_2$ are both Laves phases and are isodimorphic.

The low-temperature modifications of the compounds have a MgCu$_2$-type cubic structure. This modification exists up to 1300° in the titanium—chromium system and up to 1590° in the niobium—chromium one. The lattice parameter of TiCr$_2$ is 6.943 A, whereas that of NbCr$_2$ is 6.95 A.

The high-temperature modifications of the compounds have a hexagonal, MgZn$_2$-type structure. The lattice parameters for TiCr$_2$ are a = 4.921 A, C = 7.984 A; those for NbCr$_2$ are a = 4.92 A, c = 8.11 A.

Based on the chromium atom common to both compounds, the identical structural types, and the very slight difference in metallochemical properties between niobium and titanium, the hypothesis was advanced that the high-temperature modifications of the compounds under investigation form one continuous series of solid solutions, whereas the low-temperature modifications form another such series.

The fusibility diagram, plotted from thermal analysis data, has the form characteristic of systems with continuous series of solid solutions (Fig. 28). The temperature of polymorphic transformation found for TiCr$_2$ is 1220 ± 10°, whereas that for NbCr$_2$ is 1585 ± 10°. The temperature of polymorphic transformation increases monotonically from TiCr$_2$ to NbCr$_2$. This shows that the low- and high-temperature modifications of TiCr$_2$ and NbCr$_2$ respectively form continuous series of mutual solid solutions in the entire concentration interval. The solid solution based on the high-temperature form is called the δ-phase, whereas that based on the low-temperature form is called the γ-phase. In the binary system titanium—chromium, TiCr$_2$ is formed from solid solution with a body-centered cubic structure (β-phase). This explains the appearance in the phase diagram (see Fig. 28) of the one-phase region β, the two-phase regions β + δ and β + L, and the three-phase region L + β + δ, which extends up to 30% NbCr$_2$. No additional phase regions are observed on the NbCr$_2$ side in this system owing to the congruent melting of NbCr$_2$.

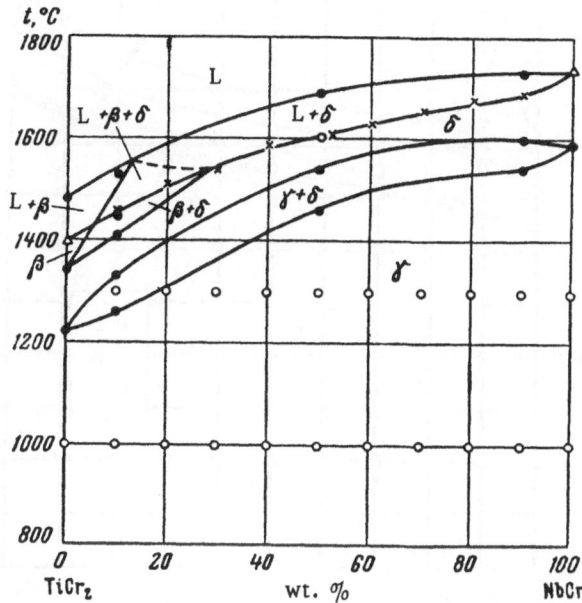

Fig. 28. Polythermic equilibrium diagram of system TiCr—NbCr$_2$: ● — high-temperature thermal analysis; × — optical method; ○ — x-ray phase analysis.

The results of thermal analysis and the phase regions in the equilibrium diagram of the studied system were confirmed by x-ray analysis. Alloys quenched from 1300° showed that at this temperature TiCr$_2$ exists in the hexagonal form. The alloy containing 90% TiCr$_2$ + 10% NbCr$_2$ gives two systems of lines, corresponding to the hexagonal and cubic modifications of the solid solutions, on the x-ray photograph. The x-ray patterns of alloys annealed at 1000° for 200 hr show only one line system, corresponding to the cubic modification. Thus both methods gave unambiguous results and confirmed the existence of continuous solid solutions in the system TiCr$_2$—NbCr$_2$, formed by both modifications of these compounds.

As is evident from the above data, Laves phases can form continuous mutual solid solutions when the main conditions are met. The numerous compounds of this type listed in Table 6 can be combined to give a large number of binary metallide systems in which the formation of continuous solid solutions may be expected. Thus, for instance, proceeding from the isomorphism of compounds of titanium analogs, one may expect Laves phases composed of zirconium or hafnium and transition or other metals to form continuous mutual solid solutions. Systems of rare-earth metal compounds which, according to [147], form continuous solid solutions are listed in Table 20 together with a number of systems in which the formation of such solutions may be presumed.

In some of these systems one may expect a break in continuity owing to the presence of different polymorphic modifications of these compounds. This was proved experimentally in the case of the metallide system ZrCr$_2$—ZrFe$_2$, where ZrCr$_2$ has two modifications (see above) and ZrFe$_2$ has one (cubic). Owing to this there is a break in solubility in the equilibrium diagram of the given system.

Based on the results of experimental investigations and well-grounded assumptions, one may presume that there are very many systems composed of Laves phases, in which continuous solid solutions can occur. Further investigations of these phenomena are of undoubted interest.

Since we cannot give a detailed description of experimental investigations in the field of binary metallide systems based on Laves phases, we note that limited solid solutions are formed in the system TiFe$_2$—ZrFe$_2$ [272] owing to the difference in type of structure of TiFe$_2$ and ZrFe$_2$. Continuous solid solutions are formed in

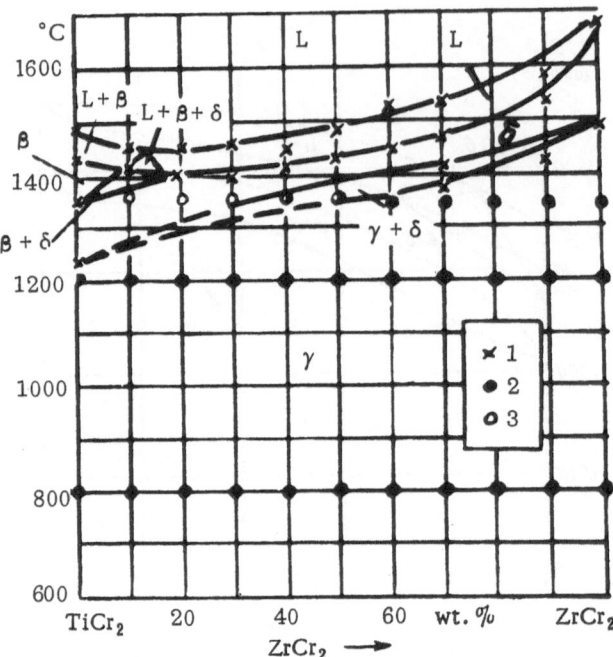

Fig. 29. Continuous solid solutions in system $TiCr_2-ZrCr_2$.

the system $ZrFe_2-ZrCo_2$ [272], since the compounds are isomorphous, and the components satisfy the rule of formation of continuous solid solutions with regard to other metallochemical properties.

An investigation of the equilibrium diagram of the system $TiCr_2-ZrCr_2$ [278], published in 1965, is of this type. Both these compounds are Laves phases; each is dimorphic: $TiCr_2$ has a hexagonal modification (type $MgZn_2-C_{14}$) at temperatures above 1220° and a cubic one (type $MgCu_2-C_{15}$) at low temperatures; $ZrCr_2$ also has a $MgZn_2$-type structure above 1480° and a $MgCu_2$- type one at low temperatures.

Thus, despite the complex crystalline state of these compounds, they are isodimorphic. In this monograph we referred the indicated system hypothetically (see Table 20 on p. 55) to systems in which total miscibility is to be expected. Actually, as a subsequent experimental investigation showed [278], continuous solid solutions are formed in this system despite its complexity.

As is evident from Fig. 29, the equilibrium diagram of this system [278] shows the formation of continuous solid solutions based on both modifications of the compounds. The equilibrium diagram at high temperatures is complicated by the fact that $TiCr_2$ is formed from the β -solid solution of the system Ti—Cr. Hence the three-phase region L + β + δ appears when alloys crystallize from the liquid phase. The rest of the diagram (Fig. 29) is self-explanatory.

Solid Solutions of $TiNi_3$-Type Metallides

Among metallides, compounds are encountered which crystallize in the hexagonal system with $TiNi_3$-type structures. The structures of such compounds have four-layer close packing, and the atoms in them are ordered [59].

These compounds are formed when metals of Groups IV (Ti, Zr, Hf) and V (Nb, Ta) react with metals in which the d electron shell is completely filled (Cu, Au) or nearly so (Ni, Pd, Pt).

Experimental data on the phase diagrams for the corresponding binary systems [43] were used to determine the compositions of many compounds having the general formula MeX_3, which may be arranged the following series: $TiNi_3$, $ZrNi_3$, $HfNi_3$, $NbNi_3$, $TaNi_3$, $TiPd_3$, $ZrPt_3$, $TiAu_3$, etc. To judge from the compositions and structures of these compounds, it may be presumed that when they interact, they will satisfy the main conditions for the formation of continuous solid solutions.

In [154, 155] the existence of continuous solid solutions in the following metallide systems was proved experimentally: $NbNi_3-TaNi_3$ [154], $TiNi_2-NbNi_3$, and $TiNi_3-TaNi_3$ [155].

The binary system $NbNi_3-TaNi_3$ is one of the first systems investigated in detail by methods of physico-chemical analysis [154]. In the case of this system it was shown that the alloys crystallize in the form of continuous metallide solid solutions and have polyhedral structure in the entire concentration range of the compounds; the electrical resistivity, hardness, and density of the alloys vary along continuous curves. The fusibility and property—composition diagrams for the system $NbNi_3-TaNi_3$, shown in Fig. 30a, confirm the presence of continuous solid solutions in it.

Subsequent investigation of the last two metallide systems ($TiNi_3-NbNi_3$ and $TiNi_3-TaNi_3$) [155] also confirmed that continuous mutual solid solutions occur in them. These conclusions are illustrated by the fusibility and property—composition diagrams shown in Fig. 30 b and c.

Recently the structures of $NbNi_3$ and $TaNi_3$ were determined more accurately; they are of the $TiCu_3$ type (orthorhombic system). The possible presence of an orthorhombic modification of $TiNi_3$ at high temperatures and a hexagonal modification at low ones may cause a break in continuous solubility in the systems $TiNi_3-NbNi_3$ and $TiNi_3-TaNi_3$.

Proceeding from the similarity in composition of the compounds and their isomorphism, one may presume that similar compounds of titanium analogs — Zr and Hg— form continuous mutual solid solutions. Such solid solutions also are possible in the systems $TiNi_3-ZrNi_3$, $TiNi_3-HfNi_3$, $ZrNi_3-HfNi_3$, etc.

In Table 21 is given a list of metallide systems with $TiNi_3$- type structures, in which continuous solid solutions are formed or should be expected.

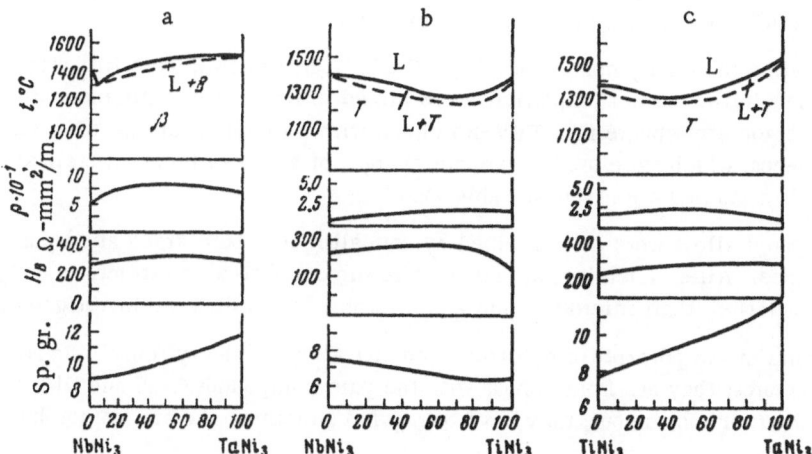

Fig. 30. Property—composition diagrams for continuous solid solutions in metallide systems: a) $NbNi_3-TaNi_3$; b) $TiNi_3-NbNi_3$; c) $TaNi_3-TiNi_3$.

TABLE 21   Systems of Continuous Solid Solutions of Metallides with $TiNi_3$-Type
Structures

Metallide system	Literature	Metallide system	Literature
$NbNi_3$ — $TaNi_3$	[154]	$TiNi_3$ — $TiPd_3$	Hypoth.
$TiNi_3$ — $NbNi_3$	[155]	$TiNi_3$ — $TiPt_3$	»
$TiNi_3$ — $TaNi_3$	[155]	$TiNi_3$ — $ZrPd_3$	»
$TiNi_3$ — $ZrNi_3$	Hypoth.	$TiNi_3$ — $ZrPt_3$	»
$TiNi_3$ — $HfNi_3$	»	$TiCu_3$ — $TiAu_3$	»
$ZrNi_3$ — $HfNi_3$	»		

Solid Solutions of Borides

As was noted in [53, 54], many borides crystallize isostructurally. Borides are exceptionally varied in structure; they include five systems and more than 15 structure types. Compounds with $CuAl_2$-, TaB-, and Al_2B-type structures are encountered most often among borides. When certain conditions are met, metal borides also can give continuous mutual solid solutions [118-120]. The tendency toward isomorphous replacement is especially marked among transition metal borides, which include borides of metals in Groups IV, V, VI, and VIII of the periodic system.

Many boride systems have been studied experimentally, and the existence of continuous solid solutions in a large number of them has been proved. Some experimentally studied boride systems, corresponding to the compositions Me_3B, Me_2B, MeB, and MeB_2, will be considered below as examples.

Of such systems, one should point out the two metallide systems Ni_3B—Co_3B and Ni_2B—Co_2B, studied while investigating the phase diagram of the ternary system Ni—C—B [156]. Contrary to literature data [43] on the binary systems Ni–B and Co–B, it was proved in [157] that the isomorphous compounds Ni_3B and Co_3B, crystallizing with rhombic (Fe_3C-type) structures, are formed in these systems, and the tetragonal compounds Ni_2B and Co_2B, also isomorphous in structure ($CuAl_2$-type), which were reported earlier [43], were confirmed.

The presence of pairs of isomorphous nickel and cobalt borides made possible the formation of continuous solid solutions by the indicated compounds. In connection with this, two metallide systems were studied experimentally: Co_2B—Ni_3B and Co_2B—Ni_2B, which were described in [156].

For the investigation the binary compounds Co_3B, Co_2B, Ni_3B, and Ni_2B were first prepared. For this, metal and boron powders in the correct proportions were mixed intimately, the mixtures pressed, and the pressings melted in a pure argon atmosphere in a TVV-2 furnace with a molybdenum heater. The cast samples were broken into a small pieces, which were used to prepare charges of alloys of binary metallide systems. The compositions of the alloys studied are given in Table 22.

The freshly prepared alloys were homogenized by annealing in a pure argon atmosphere at 1000° for 48 hr and cooled in the furnace. After annealing, all alloys were subjected to x-ray structure analysis by the powder method using cobalt radiation; their microstructure and microhardness also were investigated.

The x-ray patterns of the prepared samples revealed that alloys of the system Co_2B—Ni_3B are single-phase and have rhombic structures; they are isomorphous with the pure compounds Co_3B and Ni_3B. Alloys of the system Co_2B—Ni_2B also are homogeneous; they have tetragonal structures and are isomorphous with the compounds Co_2B and Ni_2B.

X-ray patterns of the borides Co_3B and Ni_3B and the alloy of this system containing 50% Co_3B are shown in Fig. 31, whereas x-ray patterns of the borides Co_2B and Ni_2B and the alloy containing 50% Co_2B are given in Fig. 32. In each of these figures the x-ray patterns show identical lines, corresponding to a rhombic lattice in the first system and a tetragonal one in the second.

TABLE 22. Alloy Compositions (Wt. %) in the Systems Co_2B-Ni_3B and
Co_2B-Ni_2B

System Co_3B-Ni_3B			System Co_2B-Ni_2B		
Alloy No.	Co_3B	Ni_3B	Alloy No.	Co_2B	Ni_2B
1	10	90	1	20	80
2	20	80	2	40	60
3	30	70	3	50	50
4	40	60	4	60	40
5	50	50	5	80	20
6	60	40	6	100	—
7	70	30			
8	80	20			
9	90	10			

For both systems under consideration, the unit cell volume decreases steadily as the composition of the metallide solid solution varies from the pure cobalt boride to the pure nickel boride.

Microscopic analysis confirmed the conclusion that homogeneous structures are formed in both systems. Microhardness measurements of alloys of the system Co_2B-Ni_3B showed that the hardness of continuous solid solutions in this sytem is practically independent of alloy composition and amounts to about 1145 kg/mm^2 under a

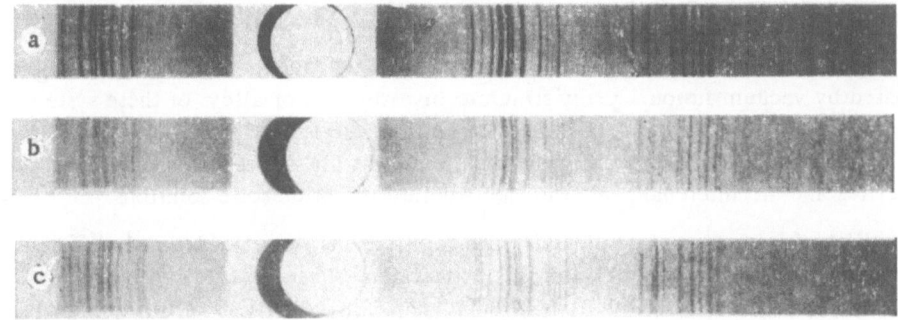

Fig. 31. X-ray patterns of alloys of system Co_3B-Ni_3B: a) Co_3B; b) 50%Co_3B; c) Ni_3B.

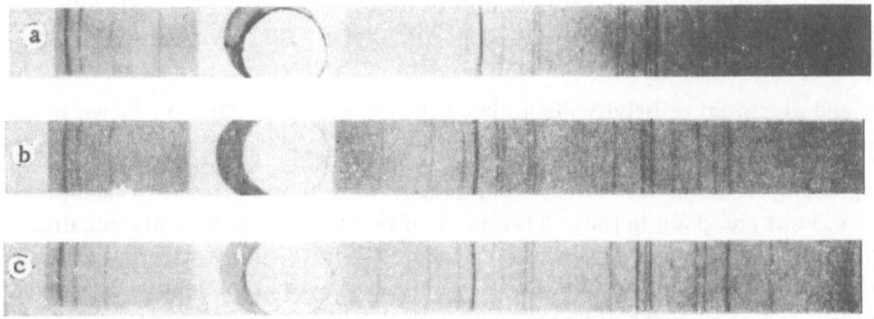

Fig. 32. X-ray patterns of alloys of system Co_2B-Ni_2B: a) Co_2B; b) 50%Co_2B; c) Ni_2B.

TABLE 23. Systems with Continuous Solid Solutions of Transition Metal Borides

Boride system	Literature	Boride system	Literature
$Ni_3B - Co_3B$	[157]	$Nb_3B_4 - Ta_3B_4$	[53, 118—120]
$Ti_2B - Ta_2B$	[53, 111, 118—120]	$TiB_2 - ZrB_2$	Hypoth.
$Cr_2B - Ti_2B$	Hypoth.	$ZrB_2 - HfB_2$	»
$Mo_2B - Ti_2B$	»	$NbB_2 - VB_2$	»
$Mo_2B - Cr_2B$	»	$VB_2 - TiB_2$	»
$W_2B - Ta_2B$	»	$TaB_2 - TiB_2$	»
$W_2B - Cr_2B$	»	$CrB_2 - VB_2$	»
$W_2B - Mo_2B$	»	$TaB_2 - NbB_2$	»
$Fe_2B - Mn_2B$	»	$TaB_2 - ZrB_2$	»
$Co_2B - Mn_2B$	»	$TaB_2 - HfB_2$	»
$Co_2B - Fe_2B$	»	$TaB_2 - VB_2$	»
$Ni_2B - Mn_2B$	»	$CrB_2 - TiB_2$	»
$Ni_2B - Fe_2B$	»	$CrB_2 - MoB_2$	»
$Ni_2B - Co_2B$	[157]	$MoB_2 - TiB_2$	»
$NbB - VB$	[53, 118—120]	$MoB_2 - VB_2$	»
$TaB - VB$	Hypoth.	$MoB_2 - NbB_2$	»
$TaB - NbB$	»	$Mo_2B_5 - W_2B_5$	»
$CrB - VB$	»	$CeB_6 - LaB_6$	[160]
$MoB - VB$	»	$CeB_6 - YB_6$	[53]
$MoB - NbB$	»	$LaB_6 - YB_6$	»
$VB - WB$	»		

50-g load; this is the same as for the initial compounds Co_3B and Ni_3B. This fact is due to the close similarity in metallochemical properties of analogous metals (Ni and Co), which lie side by side in the periodic system.

The analogous boride systems $Fe_2B - Mn_2B$, $Co_2B - Mn_2B$, and $Co_2B - Fe_2B$ were investigated in [158]. The alloys were prepared by vacuum fusion. X-ray structure investigation of alloys of these systems after homogenization annealing showed that continuous series of solid solutions are formed in all of them. This follows from the fact that they have a common atom (boron), whereas the metals Mn, Fe, and Co have nearly identical metallochemical properties and are interchangeable in the structure of boride solid solutions.

The possibility of formation of continuous solid solutions between borides with different atomic ratios can be shown and some experimental data cited in this connection. Thus in [159] the system $CrB - MoB$ was studied and the formation of continuous solid solutions in it proved. The conditions of formation of such solid solutions between monoborides are met by the systems $NbB - VB$, $NbB - TaB$, $VB - TaB$, $CrB - VB$, $VB - MoB$, $MoB - NbB$, $WB - NbB$, and many others, as will be shown below (Table 23).

An especially large number of systems consisting of transition metal diborides have been studied experimentally. In [53, 118-120] the studied systems $TiB_2 - ZrB_2$ [118], $ZrB_2 - HfB_2$ [119], $TiB_2 - HfB_2$ [119], $VB_2 - NbB_2$ [58], and many others are described. In all diboride systems satisfying the main conditions of formation of continuous solid solutions, such solutions are formed.

Fusibility and electrical resistivity diagrams of the system $TiB_2 - ZrB_2$ are shown in Fig. 33. As is evident, these two diborides, formed by two analogous metals (Ti and Zr), are mutually soluble in the solid state.

Interesting examples of the formation of continuous solid solutions between hexaborides of rare-earth metals (La, Ce, Y, etc.) are given in [53]. This group of compounds has recently acquired considerable interest in connection with their high emissivity and electrical resistivity. An investigation of the compounds LaB_6, CeB_6, and one alloy containing 50% of each by structural and physical methods [160] proved that the solid solutions formed by these analogous hexaborides are continuous. Solid solutions of the same kind were found [53] in the systems $CeB_6 - YB_6$ and $LaB_6 - YB_6$.

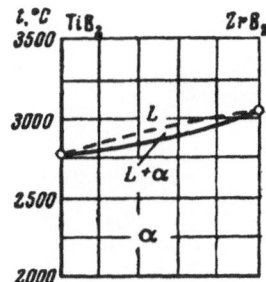

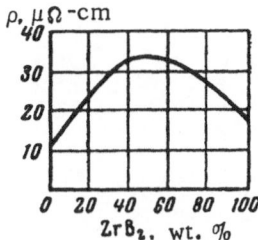

Fig. 33. Diagrams of fusibility and electrical resistivity versus composition for the system $TiB_2 - ZrB_2$.

Some boride systems, in which the formation of continuous boride solutions either has been established experimentally or may be presumed, are listed in Table 23. This composite table was compiled from data of [53, 111-113, 118-120] and includes many systems of the borides given in Table 8 of this book.

All these systems, composed of borides of analogous metals or metals with nearly identical metallochemical properties, satisfy the main conditions of formation of continuous metallide solid solutions. Experimental investigations of such systems confirm these conclusions.

Solid Solutions of Carbides

Carbides, similarly to borides, tend to form continuous mutual solid solutions. This is especially characteristic of transition metal carbides. The mutual solubility of such carbides is due to the fact that their compositions satisfy the main conditions for the formation of solid solutions between them. Many carbides contain analogous metals or metals with nearly identical metallochemical properties. Continuous solid solutions of carbides are formed between compounds having the general formula Me_3C, Me_2C, MeC, or some other composition.

Many such systems were investigated, mainly during studies of $Me_1 - Me_2 - C$ equilibrium diagrams for the so-called quasi-binary systems of carbides. They are described in the above-mentioned monographs [54,78] and article [131].

The phase diagram of the ternary system $Ti - Ta - C$, in which the quasi-binary section $TiC - TaC$ was studied [161], is shown in Fig. 34 as an illustration. It is evident from this diagram that the section $TiC - TaC$ corresponds to continuous solid solutions and that the region of metallide solid solutions of these carbides, called the δ-phase in Fig. 34, extends toward the side of alloys rich in Ti and Ta in the concentration triangle.

A striking example of the formation of continuous solid solutions between monocarbides is the system $VC - NbC$. Diagrams of fusibility and microhardness versus composition for this system are shown in Fig. 35; it follows from them that alloys of the system crystallize in the form of solid solutions, and the microhardness-composition curve has a flat maximum. Many analogous carbide systems with continuous solid solutions may be cited. Here we shall limit ourselves to a composite table, where many studied and hypothetical carbide systems are given in which continuous solid solutions exist or may be presumed (Table 24).

As is evident from Table 24, some of the systems listed there consist of metastable carbides of iron-group metals, having the general formula Me_3C. Some of these systems have been studied, whereas solid solutions may be presumed to exist in the others. For instance, solid solutions apparently can be formed in such systems as $V_2C - Nb_2C$, $V_2C - Ta_2C$, and $Nb_2C - Ta_2C$. These three carbides of analogous metals are isomorphous and have hexagonal structures with nearly identical lattice parameters. Mo_2C and W_2C, which also have hexagonal structures and apparently can form solid solutions between themselves and with V, Nb, and Ta carbides, should be referred to this group of isomorphous carbides. Most of the systems consist of transition metal monocarbides; with regard to composition and structure, they satisfy the main conditions of formation of continuous metallide solid solutions most completely.

In [279] the preparation of solid solutions of hafnium carbide with titanium, zirconium, niobium, and tantalum carbides was investigated and their physical properties studied. The authors examined in detail alloys with hafnium carbide in the binary systems: $HfC - TiC$, $HfC - ZrC$, $HfC - NbC$, and $HfC - TaC$.

Structure investigations proved that continuous carbide solid solutions are formed in all the above systems.

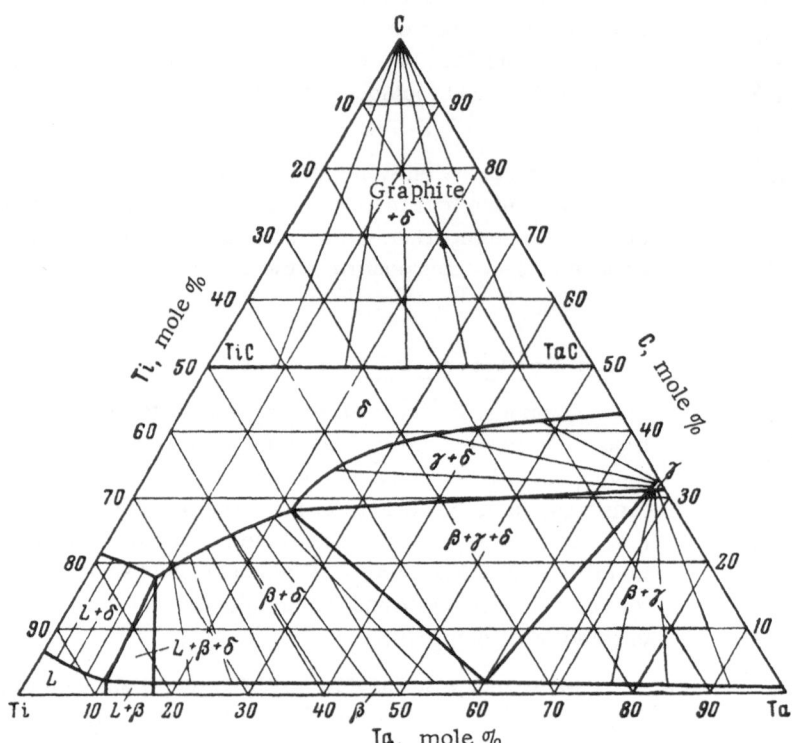

Fig. 34. Phase diagram of ternary system Ti−Ta−C with quasi-binary
section TiC−TaC at 1820°.

TABLE 24. Systems of Continuous Carbide Solid Solutions

System	Literature	System	Literature
$Fe_3C — Mn_3C$	[162]	$ZrC — HfC$	Hypoth.
$Fe_3C — Ni_3C$	Hypoth.	$TiC — VC$	»
$Fe_3C — Co_3C$	»	$TiC — NbC$	»
$Ni_3C — Mn_3C$	»	$TiC — TaC$	»
$Ni_3C — Co_3C$	»	$ZrC — NbC$	»
$Co_3C — Mn_3C$	»	$ZrC — TaC$	»
$V_2C — Nb_2C$	»	$UC — NbC$	»
$V_2C — Ta_2C$	»	$UC — TaC$	»
$V_2C — Mo_2C$	»	$UC — ZrC$	[54, 78, 111, 163]
$V_2C — W_2C$	»	$HfC — NbC$	Hypoth.
$Nb_2C — Ta_2C$	»	$HfC — TaC$	»
$Nb_2C — Mo_2C$	»	$ThC — UC$	»
$Nb_2C — W_2C$	»	$VC — TaC$	»
$Mo_2C — W_2C$	»	$NbC — TaC$	»
$TiC — ZrC$	[54, 78, 111, 163]	$NbC —UC$	»
$Sc — TiC$	Hypoth.	$TaC — UC$	»
$TiC — HfC$	»	$MoC — WC$	»

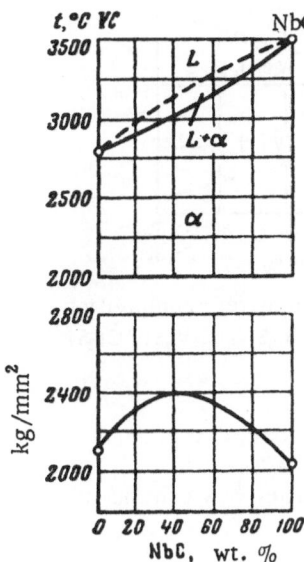

Fig. 35. Equilibrium and microhardness—composition diagrams of system VC—NbC.

For some systems the curves of incipient melting point (for the system HfC—ZrC), lattice parameter (HfC—TaC, HfC—ZrC), micrchardness, electrical, and other properties versus composition were taken. It is very curious that on some property—composition diagrams of the system HfC—TaC, property extrema appear at the composition 25 mole % HfC. The same sort of anomaly — a microhardness minimum — was found by the authors in the region 35-50 mole % ZrC in the system HfC—ZrC. This may attest to the decomposition of solid solutions in these systems and formation of ternary Kurnakov compounds having the compositions (HfTa)₃C in the system HfC—TaC and (HfZr)C in the system HfC—ZrC. The authors of [279] also assume the occurence of such ordering reactions and ascribe some of the features of chemical individuals to these compositions. However, these questions require detailed investigation. Interesting questions regarding the increased strength of chemical bonding in solid solutions of the system HfC—ZrC, based on data on the decreased coefficient of thermal expansion of alloys of this system as compared with individual carbides, are also discussed.

Data on the superconductivity properties of alloys of the carbide systems studied, particularly the change in critical temperature of transition to the superconducting state, also are given in this article.

Solid Solutions of Silicides

Among silicides, as among borides and carbides, there are many compounds with isomorphous structures and equal atomic ratios. Cases where transition metals react with silicon are especially favorable for the formation of silicide solid solutions. Many of the silicides listed in Table 10 can interact to form continuous solid solutions when the main conditions are met. Continuous solid solutions may be formed between silicides with the general formulas Me_3Si, Me_5Si_3, $MeSi$, $MeSi_2$, or other atomic ratios. In this case one should bear in mind that many silicides with different atomic ratios, as well as some isostructural silicides, have different types of chemical bonding, the metallicity of bonding decreasing from Me_3Si to $MeSi_2$. One should take this into account when considering solid solution formation between silicides.

In the literature there are data on silicide systems in which continuous solid solutions are formed. A review of these investigations is given in the well-known monograph on silicides [55] and the original papers [163-171]. These systems include the following: Cr_3Si—$MoSi$, Fe_3Si—Mn_3Si, V_3Si—Mo_3Si, V_5Si_3—Mo_5Si_3, V_5Si_3—Nb_5Si_3, $MoSi_2$—WSi_2, $NbSi_2$—VSi_2, and many others.

An example of the formation of such solid solutions is the system Cr_3Si—Mo_3Si, which many authors have studied [55, 165, 169]. Both these compounds have cubic, β-W-type structures. The curve of electrical conductivity versus composition for this system is shown in Fig. 36. This curve, having a flat minimum, characterizes the crystallization of alloys in the form of continuous solid solutions. Solid solutions are formed analogously between isomorphous compounds in the systems Fe_3Si—Mn_3Si, V_3Si—Mo_3Si, and V_2Si—Co_3Si; this is confirmed by structure analysis data.

Cubic lattice parameter-composition curves for the systems of solid solutions V_3Si—Cr_3Si and V_3Si—Mo_3Si are shown in Fig. 37. Evidently the parameters vary linearly here.

There are a number of silicide systems in which solid solutions are formed composed of transition metal monosilicides and disilicides.

A new example of the formation of continuous solid solutions is the system $MoSi_2$—$ReSi_2$. Both silicides have tetragonal, $MoSi_2$-type crystal structures. A property—composition diagram, plotted from V. S. Neshpor's data,* is given in Fig. 38. It shows the curves of microhardness H_μ, log ρ, and the lattice parameters a and c

---

*The author thanks V. S. Neshpor for the property diagram of this system, which he supplied.

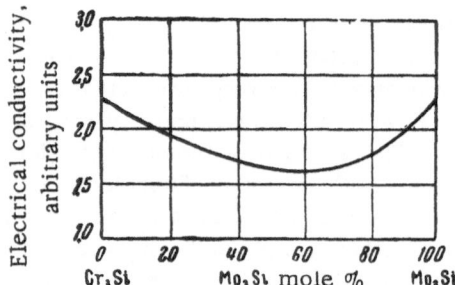

Fig. 36. Electrical conductivity−composition diagram for system Cr_3Si-Mo_3Si.

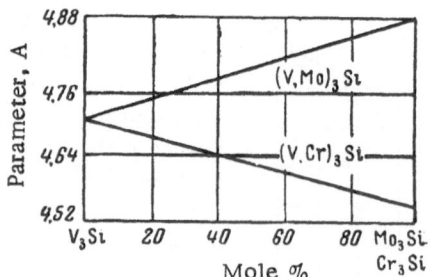

Fig. 37. Cubic lattice parameter−composition curves for systems V_3Si-Mo_3Si and V_3Si-Cr_3Si at 1300°.

versus composition for alloys of this system. All these curves indicate that the properties vary continuously, which corresponds to continuous solid solutions in this system.

Rather than dwell on description of other silicide systems in which continuous solid solutions are formed, we give a list of these systems with literature sources or the designation h y p o t h e t i c a l (Table 25).

Table 25 includes a number of systems of rare-earth metal silicides in which, in our opinion, exchange is possible. This possibility is a consequence of the chemical analogy of the elements named and the structural isomorphism of the silicides, as follows from [172].

According to Table 25, some silicide systems (CrSi−MnSi, CrSi−CoSi) form continuous solid solutions even when the constituent metals of the metallides do not.

Solid Solutions of Nitrides

Nitrides are interstitial phases; most of them have metallic bonding. Many transition metal nitrides are isostructural and have various atomic ratios. These ratios are simple, being expressed by the formulas Me_3N, Me_2N, and MeN. All this favors the formation of continuous mutual solid solutions. Many nitride systems are known in the literature, in which total mutual solubility in the solid state is observed [54, 78]. One such example is a nitride system containing two analogous metals, TiN−ZrN. The fusibility and lattice parameter−composition diagrams of this system are shown in Fig. 39; they attest to the formation of continuous solid solutions in it. Data also exist on the total mutual solubility of all the mononitrides of metals in Groups IV and V except tantalum nitride. The latter, in contrast to the cubic nitrides of the other metals in Groups IV and V, crystallizes with a hexagonal structure, and hence only limited solubility is observed in its interaction with cubic nitrides. In the other nitride systems (TiN−NbN, TiN−VN, NbN−ZrN) continuous solid solutions are formed, as the nearly linear variation of their lattice parameters attests (Fig. 40). In systems based on nitrides, it is interesting that they can give solid solutions not only between themselves, but also with carbides and even lower oxides. This occurs when the main conditions for the formation of continuous metallide solid solutions are met in the interacting

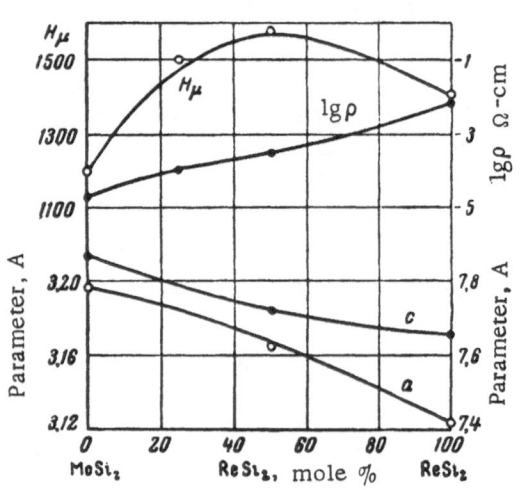

Fig. 38. Property−composition curves for system $MoSi_2-ReSi_2$.

TABLE 25. Systems of Continuous Silicide Solid Solutions

System	Literature	System	Literature
V_3Si — Cr_3Si	[55]	FeSi — VSi	Hypoth.
V_3Si — Mo_3Si	Hypoth.	VSi_2 — $CrSi_2$	[170]
Cr_3Si — Mo_3Si	»	VSi_2 — $NbSi_2$	[168]
Mn_3Si — Fe_3Si	»	VSi_2 — $TaSi_2$	[168]
Ti_5Si_3 — Zr_5Si_3	»	$NbSi_2$ — $TaSi_2$	[172]
V_5Si_3 — Cr_5Si_3	»	$MoSi_2$ — WSi_2	[168]
V_5Si_3 — Zr_5Si_3	»	USi_2 — $ThSi_2$	[171]
V_5Si_3 — Nb_5Si_3	»	$LaSi_2$ — $\beta\text{-}VSi_2$	Hypoth.
V_5Si_3 — Mo_5Si_3	[170]	$LaSi_2$ — $\beta\text{-}PrSi_2$	»
Mn_5Si_3 — Fe_5Si_3	Hypoth.	$LaSi_2$ — $\beta\text{-}SmSi_2$	»
CrSi — MnSi	[170]	$LaSi_2$ — $EuSi_2$	»
CrSi — FeSi	Hypoth	α — YSi_2 — $CeSi_2$	»
CrSi — CoSi	»	α — YSi_2 — $\alpha\text{-}PrSi_2$	»
MnSi — FeSi	»	α — YSi_2 — $\alpha\text{-}NdSi_2$	»
FeSi —CoSi	»	α — YSi_2 — $\alpha\text{-}GdSi_2$	»

systems. In the given case, where there is a common metal or metals with nearly identical metallochemical properties in nitrides and carbides or nitrides and oxides, the N and C or N and O atoms, having small but nearly identical radii, are obviously interchangeable in the isomorphous structures of the nitrides, carbides, and lower oxides. Moreover, it is important for this that these compounds have the same type of chemical bonding. Thus, based on literature data [54, 78, 172], we hold that continuous solid solutions can be formed in the systems TiN—TiC, TiN—NbC, TiN—VC, ZrN—TiC, ZrN—HfC, and many other carbide—nitride systems of transition metals. Total mutual solubility in these systems is confirmed by the curves of cubic lattice parameters for systems of these metallides, shown in Fig. 41, where their continuous character attests to the continuity of solid solutions in these systems. Developing these ideas, one may say that nitrides and lower oxides exhibit analogous mutual solubility, as is proved by the case of the TiN—TiO [172-173] and other systems. If the presence of solid solutions in titanium nitride—titanium oxide systems is confirmed, this will prove indirectly that these compounds have the same type of covalent-metallic bonding.

The compounds Ti_3O (recently found by us in the system Ti—O) and Ti_3N (found earlier [43]) are isomorphous. On our hypothesis one would expect the possible formation of continuous solid solutions between Ti_3O and Ti_3N in this case, also.

In conclusion, we give in Table 26 a number of systems based on nitrides, in which continuous solid solutions are known or may be presumed to exist.

Further investigations of systems not yet studied should confirm our hypotheses.

Solid Solutions of Superconductor Compounds

Investigations of solid solutions based on superconductor compounds are of considerable interest, in that their results lead to the discovery of more and more new materials with high superconductivity parameters. They are interesting also from the theoretical viewpoint — the development of superconducitivity theory and establishment of the scientific basis for research on new superconductive materials.

These investigations are important also from the viewpoint of determining just how the critical temperatures and magnetic fields vary with component concentration in such solid solutions. Hence, despite the fact that the study of superconductive materials is the newest field of the physics and chemistry of metals, questions

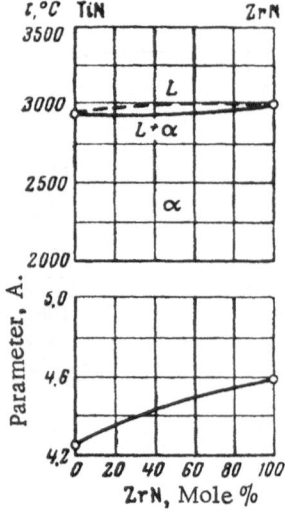

Fig. 39. Equilibrium and lattice parameter–composition diagrams of system TiN–ZrN.

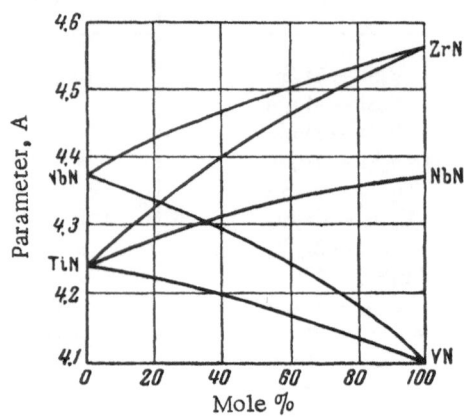

Fig. 40. Lattice parameters of nitride systems NbN–ZrN, TiN–ZrN, TiN–NbN, TiN–VN and NbN–VN.

of solid solution formation between superconductor compounds have recently been investigated extensively. In this connection, close contact between physical and physicochemical investigations of the problem of superconductivity and superconductive materials should play a large part. Certain rules established in metal chemistry, in the study of chemical reactions between metallic compounds, should be very important for recognizing questions of metallide solid solutions and predicting the latter.

At present there are indications that new superconductor materials can be prepared from nonsuperconductive metals or compounds by the formation of solid solutions with other elements or compounds. This is proved in the case of molybdenum, where molybdenum alloys can be made superconductive through the formation of solid solutions with titanium, niobium, rhodium, rhenium, technetium, and other metals.

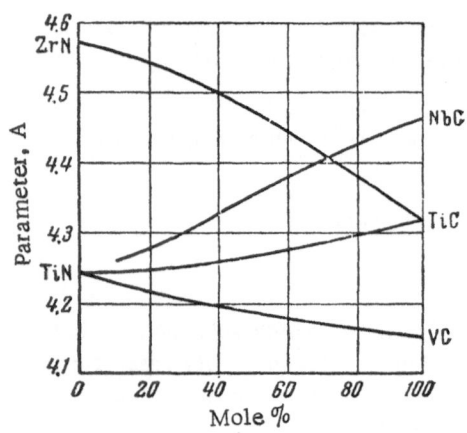

Fig. 41. Lattice parameters of nitride–carbide systems TiN–TiC, TiN–NbC, TiN–VC, ZrN–TiC.

A large number of superconductive metallides, considered above, belong to different classes of compounds, many of which satisfy the main conditions of formation of continuous mutual solid solutions. This is due to the fact that many compounds have isomorphous structures (β-W, α-Mn, and σ-phases) with one common element and either analogous metals or metals with nearly identical metallochemical properties; they have the same type of chemical bonding. All this establishes the prerequisites for mutual solubility of superconductor compounds.

Let us consider some experimental materials on solid solutions between superconductor compounds. One such paper describes an investigation of superconductivity in a number of systems consisting of transition metal carbides [174]. The systems TaC–WC, MoC–NbC, MoC–ZrC, MoC–VC, and MoC–TiC were studied; in them the relation between critical temperature and component concentration in the carbide solid solutions was determined.

TABLE 26. Systems of Continuous Solid Solutions Based on Nitrides

Metallide systems	Literature	Metallide systems	Literature
TiN — ZrN	[54, 78]	ZrN — ZrC	[54, 78]
TiN — HfN	Hypoth	ZrN — TiC	Hypoth
TiN — VN	»	VN — VC	»
TiN — NbN	»	NbN — NbC	»
ZrN — HfN	»	Nb_2N — Nb_2C	»
ZrN — NbN	»	HfN — HfO	[172]
HfN — NbN	»	TiN — TiO	[172, 173]
MoN — WN	»	ZrN — TiO	Hypoth
VN — NbN	»	ZrN — ZrO	»
NbN — TaN	»	HfN — HfO	»
TiN — TiC	»	Ti_3N — Ti_3O	»
TiN — NbC	»	Ti_3N — Ti_3C ?	»
TiN — VC	»	Ti_3N — Zr_3O	»

A diagram of the critical temperature of the system TaC—WC versus content of WC in TaC, plotted from data of [174], is given in Fig. 42. The number of valence electrons per atom e/A in this sytem is shown at the top of the diagram. Although these two carbides crystallize with different structures (TaC—cub.; WC—hexag.), T_c in this system varies along a smooth curve with a flat maximum. The highest values of T_c correspond to concentrations of 50-70% WC and an empirical ratio e/A = 4.7-4.8. In the opinion of the authors [174], the hexagonal modification of WC suppresses the cubic structure of TaC; this accounts for the nearly continuous T_c-composition curve for this system. The curves of T_c versus concentration of NbC, TiC, ZrC, and VC in molybdenum carbide, shown in Fig. 43, were based, according to [174], on the hypothetical α-modification of MoC with a cubic stru ture, rather than the stable γ-MoC, which has a hexagonal structure.

Proceeding from this cubic modification of MoC, one may presume that the systems MoC—NbC, MoC—TiC, MoC—ZrC, and MoC—VC should give continuous solid solutions. The corresponding T_c-composition curves for the systems, shown in Fig. 43, are smooth. In the system MoC—NbC the T_c curve has a flat maximum, reaching 12.5°K at ~ 50 %NbC; in the other systems T_c decreases gradually with increasing concentration of the carbides TiC, ZrC, and VC in MoC. Thus, as might have been expected, the curves of critical temperature of transition to the superconducting state versus composition for such systems of metallide solid solutions are smooth.

In systematic investigations conducted by G. V. Samsonov, N. E. Alekseevskii, and co-workers,* critical temperatures were determined in several carbide systems: TiC—VC, ZrC—NbC, and HfC-TaC, which consist of continuous solid solutions. In these systems the T_c-composition curves are complex, showing maxima and minima whose nature is not clear as yet. These authors established superconductivity phenomena in certain systems MeN—MeC, where T_c varies with composition. There are also indications [90] that T_c varies similarly in continuous solid solutions of the system NbN—NbC.

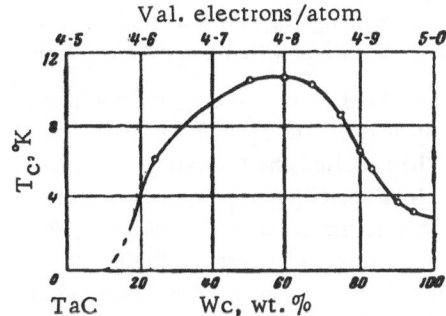

Fig. 42. Critical temperature T_c of transition to superconducting state in system TaC—WC.

*Author's report to the Conference on Physicochemical Properties and Methods of Preparation of Refractory Metals and Compounds. Kiev, Acad. Sci. Ukr. SSR, April, 1963.

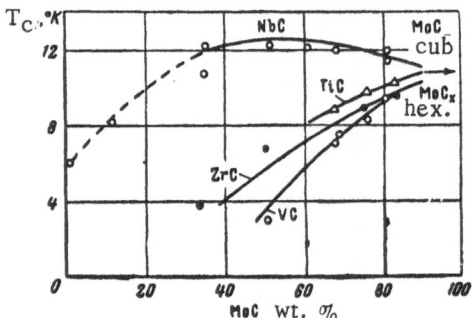

Fig. 43. Critical temperature T_C of transition to superconducting state in systems MoC—NbC, MoC—TiC, MoC—ZrC, and MoC—VC.

Some rules governing the variation of critical temperature were established in the system GdRu$_2$—ThRu$_2$ [175]. In this case the compound ThRu$_2$ is a superconductor, whereas GdRu$_2$ is not.

The authors of [175], proceeding from the structural isomorphism of these compounds (Laves phases of the cubic system C15) and the nearly identical atomic radii of Gd and Th, assumed that a continuous series of solid solutions should exist in this system. This follows from their conditions of formation, which we formulated above.

In this system, as Fig. 44 shows, the critical temperature T_C of ThRu$_2$ decreases smoothly as the nonsuperconductive GdRu$_2$ is added to it. When the GdRu$_2$ content is 9-10%, it reaches zero. The point at which T_C becomes zero nearly coincides with the point where the Curie temperature θ for solid solutions of GdRu$_2$ becomes zero on addition of ThRu$_2$ (see the corresponding curves in Fig. 44). Analogous effects were found [90] in solid solutions of the systems CeRu$_2$—PrRu$_2$ and CeRu$_2$—GdRu$_2$, which have C15-type structures.

One paper on the superconductivity of metallide solid solutions deals with critical temperatures in the silicide system V$_3$Si—Mo$_3$Si [176]. In this system V$_3$Si is a superconductor with $T_C = 17°K$, whereas Mo$_3$Si has a very low critical temperature (1.3°K). In this system, however, the two silicides are isomorphous in structure (β-W); they form continuous solid solutions.

The curve of the critical temperature of V$_3$Si versus Mo$_3$Si content in the solid solution (Fig. 45) shows a considerable decrease in the former as Mo$_3$Si is added; this decrease is rather sharp up to 5% Mo and gradual above this concentration. The reasons why T_C varies in this way are obscure; further study is required.

In a series of papers on systems of superconductor compounds, an investigation of the synthesis of new niobium compounds—systems of niobium metallides with the β-W structure— is important [177]. In this study, methods were developed for preparing Nb compounds with the general formula Nb$_3$B (where B is a metal of Group IIIB, IVB, or VB: Al, Ga, In, Ge, Sn, or Sb) and alloys based on quasi-binary sections of such compounds. The metallide systems Nb$_3$Sn$_{1-x}$—Sb$_x$, Nb$_3$Sn$_{1-x}$—Al$_x$, Nb$_3$Sn$_{1-x}$—Ga$_x$, and Nb$_3$Al$_{1-x}$—Ga$_x$ were studied. The investigation established that continuous solid solutions with the β-W structure exist in the system Nb$_3$Sn—Nb$_3$Sb; in the other systems (Nb$_3$Sn—Nb$_3$Al, Nb$_3$Sn—Nb$_3$Ga, and Nb$_3$Al—Nb$_3$Ga) there is only partial mutual solubility.

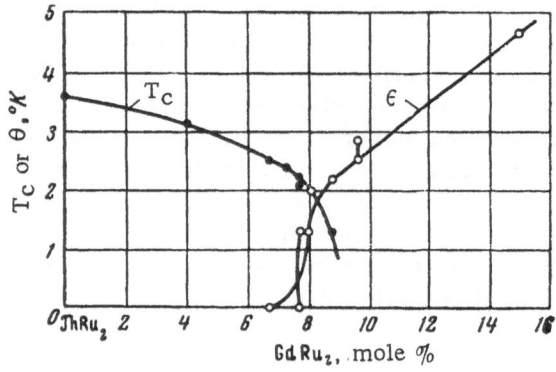

Fig. 44. Curves of critical temperature Tc and Curie point θ versus GdRu$_2$ content for system ThRu$_2$—GdRu$_2$.

The fact that total mutual solubility exists in the system Nb$_3$Sn—Nb$_3$Sb, although continuous solid solutions do not occur in the system Sn—Sb, is of interest from the general viewpoint of exchange of atoms of these metals in compounds. These questions will be considered below.

Solid solutions based on superconductor compounds were investigated most fully and systematically by the authors of [178]. They studied the formation of solid solutions between nickel arsenide-type compounds, including systems of bismuthides and antimonides of iron- and platinum-group metals. The authors studied the systems PtBi—PtSb, NiBi—RhBi, PdSb—PdBi, PtBi—PdBi, NiBi—NiSb, and PtBi—PtPb. These investigations were conducted along the quasi-binary sections AB—AC in the corresponding ternary systems. Continuous solid solutions

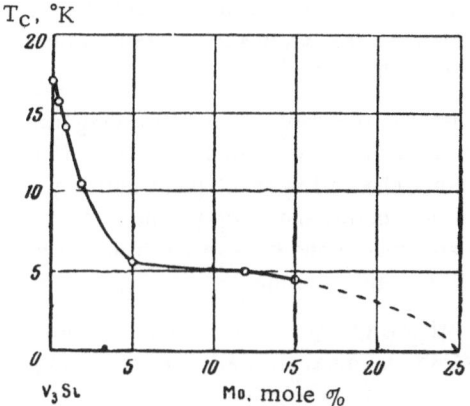

Fig. 45. Variation of critical temperature in system V_3Si-Mo_3Si.

based on superconductor compounds were found in a number of the systems studied. The results confirm the main conditions for interchangeability of metals in the formation of solid solutions based on metallic compounds, given above.

These systems of superconductor metallides were investigated by thermal analysis, microscopic and x-ray methods, and microhardness and superconductivity studies. It was found that alloys of the studied systems reach equilibrium slowly on homogenization annealing; in this case the superconductivity method is the most sensitive for detecting microheterogeneity in alloys.

The curve of average microhardness of the solid solution versus alloy composition is nearly linear in the cases of the sections PtBi—PtSb, NiBi—RhBi, and PdSb—PdBi but is nonlinear for the sections PtBi—PdBi, NiBi—NiSb, and PtBi—PtPb.

An x-ray investigation of solid solutions showed that the lattice parameters vary linearly with alloy composition within the limits of experimental error. This type of dependence of the lattice parameter of a solid solution shows that continuous solid solutions are formed in a number of the systems studied.

When superconductivity was investigated in alloys of the system PtBi—PtSb, it was shown that the critical temperature does not vary linearly with composition, even though the period and interatomic distance do. With changing alloy composition, its value does not pass through those for the superconductive components.

The results of the authors' investigation are given in Table 27. It shows that continuous solid solutions are formed in four of the twelve binary systems studied. These are systems made up of isomorphous compounds

TABLE 27. Mutual Solubility of Superconductor Compounds

Investigated systems AB—AC	Solubility in compound, mole %		Difference between atomic radii B and C	Formation of CSS in binary systems B—C
	I	II		
PtBi — PtSb	CSS	CSS	11.5	CSS
PtBi — PtPb	CSS	CSS	3.8	—
NiBi — NiSb	CSS	CSS	11.5	CSS
NiBi — RhBi	CSS	CSS	7.5	—
PtBi — PdBi	55 *	<10	0.5	CSS
PdSb — PdBi	45 *	~ 8	11.5	CSS
NiBi — PtBi	**	**	10.0	CSS
NiBi — MnBi	**	**	4.5	—
PtBi — MnBi	**	**	5.8	—
PtBi — PtSn	<10	~20	13.2	—
PtSb — PdSb	0	0	0.5	CSS
PtSb — CoSb	0	<20	9.5	CSS

*Compounds have different crystal structures.

**We did not find more than 5% solubility on the component (compound) side after homogenizing the alloys 200 hr near the temperature of formation of the compounds NiBi and MnBi.

Explanation. I —Solubility of compound AC in compound AB; II) solubility of compound AB in compound AC; CSS — continuous series of solid solutions.

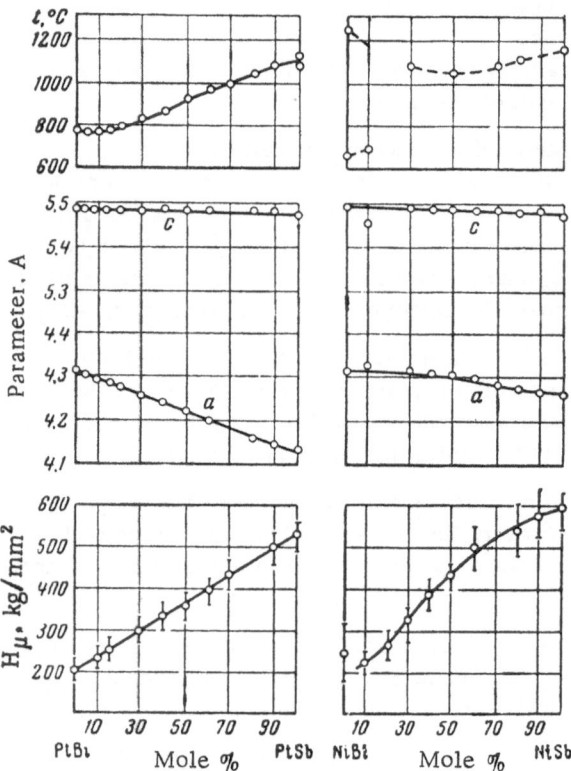

Fig. 46. Diagrams of fusibility, lattice parameter, and microhardness H_μ versus composition for solid solutions of systems PtBi—PtSb and NiBi—NiSb.

whose component atoms have similar metallochemical properties. Such are the systems PtBi—PtSb, PtBi—PtPb, NiBi—NiSb, and NiBi—RhBi.

Fusibility and property—composition diagrams are shown in Fig. 46 for the two studied systems PtBi—PtSb and NiBi—NiSb. The curves of incipient melting point, lattice parameter, and microhardness versus composition are smooth and correspond to the properties of metallide solid solutions.

In the other eight metallide systems studied, despite the close chemical similarity of their component atoms, limited solid solutions are formed. This is explained by the fact that the compounds with similar component atoms are nonisomorphous, as well as the fact that some of these compounds crystallize by a peritectic reaction with a virtual maximum.

A solubility gap in these systems is observed also owing to the absence of continuous mutual solubility of their component metals. This applies, for instance, to the systems PtBi—PtSn and PtBi—MnBi, since neither Bi and Sn nor Pt and Mn form continuous mutual solid solutions. Moreover, it should be noted that in the system PtBi—PtPb the metallides PtBi and PtPb are mutually soluble in all ratios, even though Bi and Pb do not form continuous solid solutions. Here, apparently, the adjacency of bismuth and lead in group and period affects their total exchange in the structure of solid solutions of the compounds.

Among the many experimentally established superconductor compounds listed in Table 13, one may expect the formation of a large number of continuous solid solutions. This is possible in cases where the compounds chosen for interaction satisfy the main conditions for mutual solubility of metallides. These systems include V_3Sn-Nb_3Sn, V_3Sn-Ta_3Sn, V_3Si-Nb_3Si, V_3Al-Nb_3Al, V_3Al-Mo_3Al, V_2Zr-Nb_2Zr, V_2Zr-Nb_2Ti, Mo_3Re-W_2Re, $ZrN-MoN$, Nb_3Ga-V_3Ga, $NbC-TaC$, etc. None of these systems of superconductor compounds with high critical temperatures have been studied experimentally until now; they are very interesting from both the viewpoint of establishing mutual solubility in them and that of studying the variation of superconductivity parameters with composition for systems with continuous solid solutions.

Systems of superconductor compounds, in which the formation of continuous solid solutions either has been established experimentally or is hypothetically presumed possible, are listed in Table 28. For some systems the critical temperature of each compound (in °K) is given under its formula in the table.

While this monograph was in press, a paper appeared (H. Holleck, H. Nowotny and F. Benesovsky. Monatshefte für Chemie 94, p. 359, 1963) describing an investigation of systems of metallides (superconductor) which form continuous solid solutions. The lattice parameters of the studied systems Nb_3Sn-Mo_3Al and Nb_3Sn-Ti_3Sn, the component compounds of which exhibit continuous mutual solubility, are shown in Fig. 47. This is consistent with our theses regarding the formation of continuous solid solutions between metallides.

Many of the systems given in Table 28 have not been studied with regard to interactions of metallides or the variation of superconductivity constants with composition and structure. They are of interest for further investigation.

TABLE 28. Systems of Superconductor Compounds Forming Continuous Solid Solutions

Metallide system I — II	Structure type of compound I	II	Metallide system I — II	Structure type of compound I	II
MoC — NbC 9.6 — 6	unst. cub.	cub.	PtBi — PtSb 1.21 2.1	Nickel arsen.	
MoC — ZrC 9.6 <1.2	»	»	PtBi — PtPb 1.29 —	»	
MoC — VC 9.6 <1.2	»	»	NiBi — NiSb 4.26 <1	»	
MoC — TiC 9.6 <1.2	»	»	NiBi — RhBi 1.21 2.6	»	
TaC — WC ~11 <1.8	cub.	hex.	V_3Ge — V_3Si 6.01 17	β-W	—
TiC — VC <1.20 <1.2	»	cub.	V_3Sn — Nb_3Sn 16 18.05	β-W	β-W
ZrC — NbC <1.20 6 (14)	»	»	V_3Sn — Ta_3Sn 6 6.4	β-W	β-W
HfC — TaC — ~11	»	»	V_3Si — Mo_3Si 17	β-W	β-W
NbB — VB 8.25 <1.28			Ta_3Sn — Nb_3Sn 6.4 18.05	β-W	β-W
NbB_2 — VB_2 <1.28 1.9	hex.	hex.	V_3Al — Nb_3Al — 18.1	β-W	β-W
NbN — VN 15.6 8.2	hex.	cub.	V_3Si — Mo_3Al 17	cub.	—
ZrN — MoN 8.9 12	cub.	hex.	$MoRe_3$ — WRe_3 9.89 9	α-Mn	α-Mn
TiN — VN 5.6 8.2	cub.	cub.	Nb_3Ga — V_3Ga 13.2 16.8	β-W	β-W
$GdRu_2$ — $ThRu_2$	Laves phases cub.	C15	Nb_3Sn — Nb_3Sb 18 —	β-W	β-W
MoN — NbN 12.0 12.2	hex.	hex.	Nb_3In — V_3In — —	β-W	β-W
NbN — VN 12.0 8.2	»	cub.	WRe — MoRe 5.2 6	tetrag.	tetrag.

In connection with the search for new superconductive materials among metallides and their alloys, the authors of [280] conducted a special investigation of alloys of the metallide system NbC — TaC. As was presumed, this system forms continuous solid solutions, as is evident from the smoothness of the lattice parameter—composition curve for alloys of the system (see Fig. 48a).

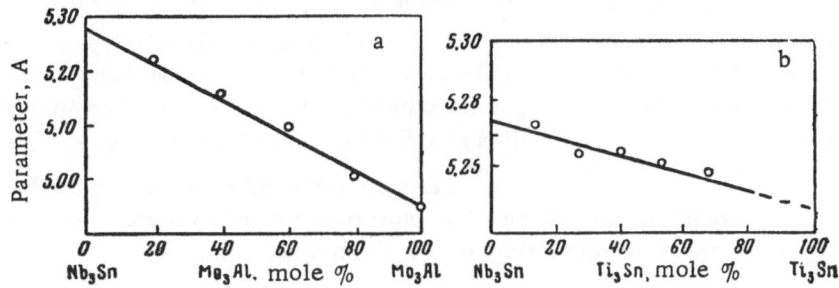

Fig. 47. Diagrams of lattice parameter versus composition for systems with continuous solid solutions: Nb_3Sn—Mo_3Al (a) and Nb_3Sn—Ti_3Sn (b).

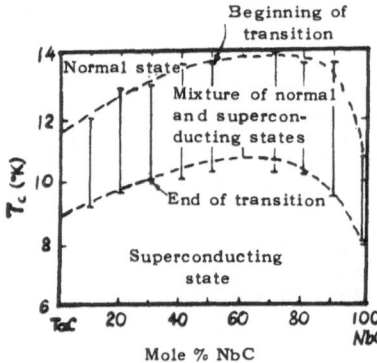 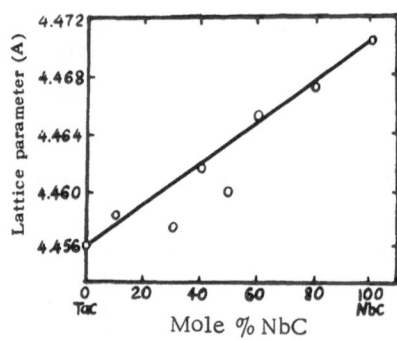

Fig. 48. a) Variation of lattice parameter in system NbC−TaC. b) Critical temperatures of beginning and end of transition to superconducting state in alloys of system NbC−TaC.

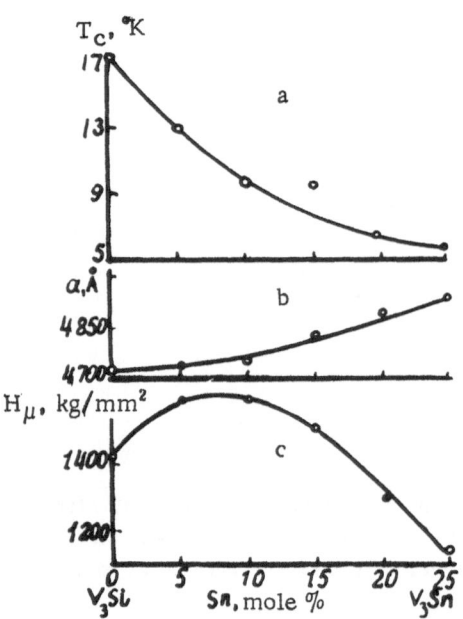

Fig. 49. a) Variation of T_C of transition to superconducting state; b) variation of lattice period; c) variation of microhardness in system of solid solutions V_3Si-V_3Sn.

Study of the critical temperature of transition of this system to the superconducting state showed that the transition takes place over a certain temperature interval. Curves of these temperature intervals versus composition are shown in Fig. 48b. As is evident from the figure, the transition temperature− composition curves for the system NbC−TaC passes through a flat maximum, contrary to some other metallide systems with superconductor properties. The authors explain the interval of transition temperatures T_C by non-correspondence of stoichiometric compositions of the carbides, chiefly with regard to their carbon content.

One recent experimental investigation of systems of superconductor compounds is [281]. The authors studied the structure and properties of alloys of the metallide system V_3Si-V_3Sn, both components of which are superconductor compounds. In this paper the formation of continuous solid solutions between these isomorphous compounds is shown by x-ray and microstructure analyses of alloys annealed for a long time (2500 hr at 800°). The transition temperature (T_C) (see Fig. 49) decreases smoothly from a maximum value (17.1°K) for V_3Si to a minimum (6.0°K) for V_3Sn. Curves of the lattice parameter and certain properties of the system studied versus composition are also shown in Fig. 49.

The lattice parameter (Cr_3Si-type structure) of the alloys varies with composition, its values being slightly less than those predicted by Vegard's law, whereas the microhardness passes through a flat maximum (see Fig. 49).

The system studied is interesting in that continuous exchange of Si and Sn atoms, whose radii and metallochemical properties are quite different, takes place in it. In the binary system Si−Sn these elements form a simple eutectic mixture.

Such examples are not unique, and the formation of continuous metallide solid solutions in sucn systems is discussed in Chapter IX of this monograph, which deals with some questions of metallide chemistry (see pp. 137−38)

Solid Solutions of Semiconductor Compounds

As was considered above, compounds with semiconductor properties are formed by the reactions of many metals with analogous nonmetals. Hence many of them have isomorphous structures, the same type of chemical bonding, and, in some cases, constituent elements with the same number of valence electrons or isoelectronic structures. All this favors the formation of continuous solid solutions between semiconductor compounds. Some important conditions for the formation of such solutions were considered in [111-113, 115-117]. Since these papers were published, many experimental investigations have been conducted in the field of interactions of semiconductor compounds, and much material has accumulated.

Many of the studied systems of semiconductor compounds exhibit total mutual solubility in the solid state; it should be noted that such solutions are formed between compounds with the most varied combinations of metal atoms with atoms of semi- and nonmetals: bismuth and antimony, arsenic and phosphorus, selenium and tellurium, etc.

These possibilities are realized when the component compounds of the system meet the main conditions of isomorphous substitution.

Such investigations are described in [115-117, 128, 179-186]. The existence of continuous solid solutions was proved in the following systems of semiconductor metallides: $GaSb-InSb$ [128], $PbTe-PbSe$ [179], $SnTe-GeTe$ [180], $InP-InAs$ [181], $GaP-GaAs$ [182], $InSb-InBi$ [184], $CdSb-InSb$ [185], $HgTe-HgSe$ [186], and many others. The investigations are generalized in a monograph [105].

It is very interesting that, as is evident from this brief list of metallide systems in which continuous solid solutions occur, these compounds are formed by the most diverse elements: Element A in the compound belongs to Group II (Hg), III (Ga, In), or IV (Pb, Ge), whereas element B belongs to Group V (P, As, Sb) or VI (Se, Te). All these elements belong to subgroup B of the periodic system.

Despite this variety of combinations of elements in such compounds, they all have covalent bonding; many of them are isomorphous and have identical atomic ratios. This explains the fact that when one group of elements A_{II}, A_{III}, A_{IV} are chemically similar, on the one hand, whereas another group B_V, B_{IV} are chemically similar, on the other, continuous solid solutions are formed in systems of such compounds. The compositions of many of these compounds conform to the laws of valence.

The two characteristic systems $GaSb-InSb$ [128] and $PbSe-PbTe$ [179], in which the formation of continuous solid solutions was shown, are given in Fig. 50 as an example.

Although the discontinuity of solid solutions in the $GaSb-InSb$ system was shown in an early paper [65], further investigations [128] using thermal and structure analyses of alloys annealed for long periods proved that continuous solubility exists in this system.

TABLE 29. Systems with Continuous Solid Solutions of Compounds Having
Semiconductor Properties

System	Literature	System	Literature
CdS — CdSe	[65]	GaP — GaAs	[65]
CdTe — InTe	[115]	AlSb — GaSb	[182]
CdTe — HgTe	[115]	InSb — InBi	[184]
HgS — HgTe	[115]	CdSb — InSb	[185]
GaSb — InSb	[128]	HgTe — HgSe	[186]
PbTe — PbSe	[179]	ZnS — CdS	[65]
SnTe — GeTe	[180]	ZnS — HgS	[65]
InP — InAs	[182]	ZnSe — HgSe	[65]
HgS — HgSe	[65]		

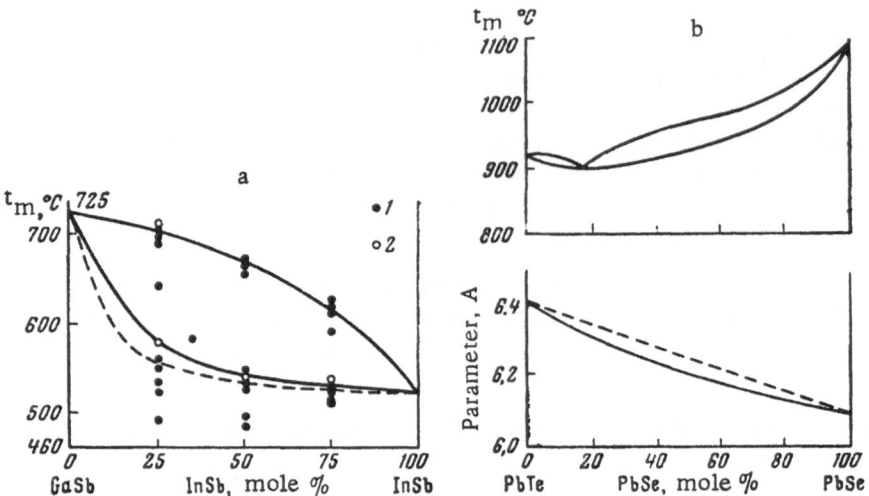

Fig. 50. Solid solutions of semiconductor-compound systems: a) GaSb−InSb;
b) PbTe−PbSe.

Some systems of semiconductor compounds AB which, as various literature data confirm, form continuous solid solutions are listed in Table 29.

As is evident from Table 29, the systems with total mutual solubility include some in which the constituent elements of the compounds do not form continuous mutual solid solutions.

Besides the above-mentioned metals Sn and Ge, forming continuous solid solutions in the SnTe−GeTe system, the element pairs Cd and In, Cd and Hg, Ga and In, Al and Ga, and P and As also should be cited from

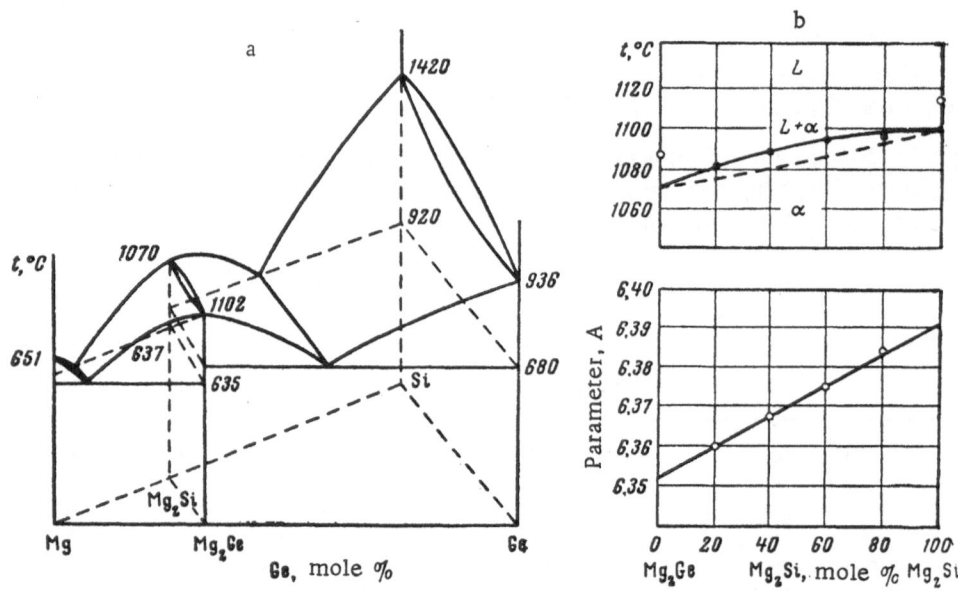

Fig. 51. System $Mg_2Si−Mg_2Ge$ with continuous solid solutions: a) fusibility diagram of ternary system Mg−Si−Ge; b) property−composition diagrams of system $Mg_2Ge−Mg_2Si$.

TABLE 30 Systems of Semiconductor Compounds with Continuous Solid Solutions

System of compounds I-II	Structure type of compounds		System of compounds I − II	Structure type of compounds	
	I	II		I	II
$A_{II}B_{IV}$-type (valence) compounds			Other sulfides, selenides, tellurides, antimonides, arsenides		
$Mg_2Si — Mg_2Ge$	cub.	cub.	$SnSe — SnTe$	cub.	cub.
$Ca_2Si — Ca_2Ge$	tetrag.	tetrag.	$As_2Se_3 — As_2Te_3$	rhomb.	—
$A_{II}B_{IV}$-type compounds			$Sb_2Se_3 — Sb_2Te_3$	»	—
$CdSe — CdTe$	cub.	cub.	$Sb_2Se_3 — As_2Se_3$	»	—
$SrSe — SrTe$	cub.	cub.	$Sb_2Se_3 — Sb_2Te_3$	»	—
$A_{III}B_{IV}$-type compounds			$Bi_2Te_3 — Bi_2Se_3$	»	—
$AlAs — AlSb$	cub.	cub.	$Bi_2Se_3 — Sb_2Te_3$	»	—
$GaAs — GaSb$	»	»	$InSe — InTe$	—	—
$InAs — InSb$	»	»	$Mg_3Sb_2 — Mg_3Bi_2$	hex.	—
			$FeAs — MnAs$	rhomb.	—

Table 29. They do not form continuous mutual solid solutions in simple metallic systems, either, but display total mutual solubility in a number of compounds. The reasons for such behavior on the part of these elements in metallide systems have not been precisely determined; they require further study and analysis. Nevertheless, one can say that either these paired elements belong to identical groups of the periodic system and are analogs (Sn and Ge, Cd and Hg, Ga and In, Ga and Al, and As and P), or they lie in adjacent groups in the same period (Cd and In). These questions will be discussed from this viewpoint below.

Using the conditions determining the formation of continuous solid solutions between metallides [111-115] and certain peculiarities of semiconductor compounds in this respect [115-117], one can compile a composite table of systems of these compounds, in which total mutual solubility in the solid state either has been established or may be expected. Such data are given in Table 30 for compounds of the types $A_{II}B_{IV}, A_{II}B_{VI}, A_{III}B_V$, etc.

One of these systems ($Mg_2Si — Mg_2Ge$) was studied recently [187]. These data confirm the formation of continuous solid solutions in it. A three-dimensional diagram of the ternary system $Si — Ge — Mg$ with the quasi-binary section $Mg_2Ge — Mg_2Si$ is shown in Fig. 51a, whereas thermal-analysis and lattice parameter curves for the system of these metallides are given in Fig. 51b.

In Table 30 we gave only those systems in which the compounds satisfy the main conditions of formation of solid solutions between metallides. As is evident from the table, the paired component compounds of each system are isomorphous (structural characteristics are lacking for some compounds). In them, metal A is the same for both (Mg and Mg, Ca and Ca, Ga and Ga, etc), whereas the role of element B is played by analogous elements which form continuous mutual solid solutions (Si and Ge, Se and Te, As and Sb, and Sb and Bi). As is well known, systems of these elements exhibit total mutual solubility.

Thus for a sound presumption of the presence of solid solutions in the systems listed in Table 30, three main conditions must be met: structural isomorphism, identity in type of bonding, and the presence in the compounds of analogous elements which can give continuous mutual solid solutions. The number of these systems may be increased by special conditions of formation of continuous solid solutions between semiconductor compounds [115-117], and also by heterovalent substitution (according to Goryunova [105].

Further investigations of these systems may confirm our hypotheses regarding the formation of continuous solid solutions between such metallides.

Other Metallide Systems with Continuous Solid Solutions

Proceeding from the validity of the above conditions for the formation of continuous metallide solid solutions, one may extend them to other possible cases of reactions between different compounds.

In addition to the above, we shall consider systems of other classes of compounds, for instance, germanides, phosphides, etc. Many compounds with metallic and covalent bonding are encountered in such systems. There are also grounds for extending these conditions to binary systems composed of ternary and more complex individual metallides. Questions of the formation of metallide solid solutions between germanides and phosphides, as well as similar compounds, have not been specially considered by anyone until now. In this connection, although existing experimental data are insufficient, total mutual solubility may be presumed in many systems. We shall give several examples of systems of various classes of compounds.

Systems Based on Germanides

Germanium, as a Group IV element, is an analog of silicon and can give a number of compounds—germanides—which are analogous to silicides, are isomorphous with them, and have identical types of chemical bonding at identical atomic ratios. In a review [188] of investigations of germanium and its compounds, particularly with regard to the conditions of formation of germanides, it is noted that germanium is closer in chemical properties and atomic structure to silicon than to gray tin. The atomic radii of germanium and silicon are nearly the same (Ge = 1.39 A, Si = 1.34 A). This explains the fact that germanium and silicon form continuous mutual solid solutions.

The most typical known germanide compounds are digermanides of the type $MeGe_2$; they are formed by the reaction of germanium with Group II metals and nearly all transition metals. Many of these compounds have skeletal structures and are isostructural with the corresponding disilicides, e.g., $CaGe_2$ and $CaSi_2$, $PrGe_2$ and $ThSi_2$. Of the peculiarities of monogermanides, one should mention isomorphous compounds, e.g., CaGe, CaSi, and CaSn, whose structures consist of zigzag chains. Among monogermanides one also should note the isomorphous compounds CrGe, ZrGe, NiGe, PdGe, etc. Of germanium compounds with a high content of electropositive metals, one should note the group of valence compounds Mg_2Ge and Ca_2Ge, isomorphous with Mg_2Si, Mg_2Sn, Mg_2Sb, and Ca_2Si, respectively. One also should note compounds with still higher metal content, such as V_3Ge, Cr_3Ge, and Mo_3Ge, which have β-W-type structures and apparently are superconductor compounds. Ni_3Ge has a cubic, $AuCu_3$-type structure, as does Ni_3Si.

There are a number of intermediate-type germanium compounds having the atomic ratio Me_3Ge_2. Among them one should note Th_3Ge_2, which is isostructural with Th_3Si_2 and U_3Si_2; compounds of germanium with metals of Groups IV-VIII, having the compositions Me_5Ge_3 and $MeGe_2$, are especially stable. All these germanides are isostructural with the corresponding silicides of these metals. Thus, for instance, $TiGe_2$ and $TiSi_2$, $ZrGe_2$ and $ZrSi_2$, $NbGe_2$ and $NbSi_2$, $MoGe_2$ and $MoSi_2$, etc., are isomorphous.

The compositions of germanium compounds with more electronegative elements are interesting. Germanium acts as a cation in such compounds, and polar forces predominate in them. They include the arsenides GeAs and $GeAs_2$, isostructural with SiAs and $SiAs_2$. In these compounds ionic bonding predominates; the sulfides GeS and GeS_2 are analogous to many oxides and are nonmetallic compounds.

The above brief list of germanides and silicides isomorphous with them shows that they satisfy the main conditions of isomorphous substitution. On this basis it may be presumed that many such compounds should give continuous mutual solid solutions. Although we do not know of any experimental data, we can, proceeding from general theses, name systems in which continuous solid solutions should exist. Such, in our opinion, are the systems: $CaGe_2-CaSi_2$, $CaGe-CaSi$, Mg_2Ge-Mg_2Si, V_3Ge-Cr_3Ge, V_3Ge-Mo_3Ge, V_3Ge-V_3Si, V_3Ge-Cr_3Si, Mo_3Ge-Mo_3Si, $Th_3Ge_2-Th_3Si_2$, $Ti_5Ge_3-Ti_5Si_3$, $TiGe_2-ZrGe_2$, $TiGe_2-TiSi_2$, $ZrGe_2-ZrSi_2$, $NbGe_2-MoGe_2$, $NbGe_2-NbSi_2$, $MoGe_2-MoSi_2$, and many others.

Further experimental investigations of such systems would be very interesting from the viewpoint of the theory of metallide interactions and the practical use of alloys based on germanides.

Systems Based on Phosphides

Based on [56], we noted that phosphides, other conditions being equal, have a smaller fraction of ionic bonding than nitrides and oxides and form compounds with metallic bonding to a larger degree than sulfur. Since it has a larger atomic radius than other nonmetals (C, N, O), phosphorus almost never gives interstitial phases with transition metals. Nevertheless, the phosphides of many transition metals are similar in structure to the carbides and nitrides; however, their unit cells are much larger owing to the larger atomic radius of phosphorus. The fact that the bonding in these compounds becomes more metallic with decreasing phosphorus content and more covalent with increasing number of phosphorus atoms is characteristic of phosphides; other metallides containing nonmetals show analogous behavior. This has been proved for a number of phosphides, e.g., the iron phosphides Fe_3P, Fe_2P, and FeP, and other compounds (see above).

The wide variety of compositions, structures, and types of chemical bonding in phosphides imposes peculiarities on their metallochemical properties and the character of interactions between them. These questions are considered in [56] and also in recent publications on phosphides and phosphorus [189, 190].

Without discussing in detail the composition and structure of phosphides having various atomic ratios, we give here some examples of the formation of compounds with the most characteristic metallic bonding. They are mainly transition metal phosphides, having the general formulas Me_3P, Me_2P, and MeP, in the compositions of which metal atoms predominate.

These compounds may be grouped according to structure and atomic ratio as follows: V_3P, Cr_3P, Mo_3P, W_3P, Mn_3P, Fe_3P, and Ni_3P — tetragonal structure; Ti_3P and Pt_3P—cubic; Cr_2P, Mn_2P, Re_2P, and Co_2P—rhombic; Fe_2P and Ni_2P— hexagonal. Among monophosphides there are also a number of isomorphous compounds of transition metals with phosphorus: TiP, β-ZrP, VP, and MoP — hexagonal structure; α- and β-NbP and α- and β-TaP — tetragonal; CrP, WP, MnP, FeP, CoP, RuP — rhombic.

The monophosphides of the Group IIIB elements boron, aluminum, gallium, and indium constitute a special group of compounds. They are all typical semiconductors with the zinc blende structure. The strength of chemical bonding in these compounds corresponds to the forbidden-gap width, which is for $BP \sim 4.5$, $AlP \sim 2.5$, $GaP \sim 2.3$, and $InP \sim 1.2$ eV. In accordance with this the fraction of ionic bonding and the strength of chemical bonding increase with forbidden-gap width, i.e., in the series $InP \to GaP \to AlP \to BP$ (see Fig. 15).

Based on the above structural peculiarities of phosphorus atoms, specifically, the large difference between their radius (see Fig. 1) and those of its analogs (N, As, Sb, and Bi) and period neighbors, one would hardly expect total exchange of its atoms with those of its analogs in solid solution formation between phosphides and other compounds.

Transition metal phosphides having identical atomic ratios also are exceptionally varied in structure (see above). The number of isostructural compounds of phosphorus with analogous metals or metals with similar metallochemical properties is limited. All this, in turn, evidently limits the number of phosphide systems that can give continuous solid solutions. These questions have received little study. Some material on quasi-binary sections of phosphides is collected in the monograph mentioned above [56]. Since the amount of experimental material available to us is not large, we give those systems in which, based on experimental data or general theoretical theses regarding the formation of continuous solid solutions between metallides, one may expect total mutual solubility of phosphides.

V_3P–Cr_3P, V_3P–Mo_3P, V_3P–W_3P, Cr_3P–Mo_3P, Cr_3P–W_3P, Mo_3P–W_3P, Cr_3P–Fe_2P [56], Mn_3P–Fe_3P, Mn_3P–Ni_3P, Fe_3P–Ni_3P [56], Fe_2P–Ni_2P [56], Fe_2P–Mn_2P [56], TiP–βZrP, TiP–VP, TiP–MoP, VP–MoP, α- and β-NbP, α- and β-TaP, FeP–CoP, etc.

There are some data [56] on the continuity of solid solutions formed by indium and arsenic monophosphides InP–AsP [56], which are semiconductor compounds, and also by the isomorphous compounds Zn_3P_2 and Cd_3P_2. These compounds and the solid solutions formed between them also have semiconductor properties.

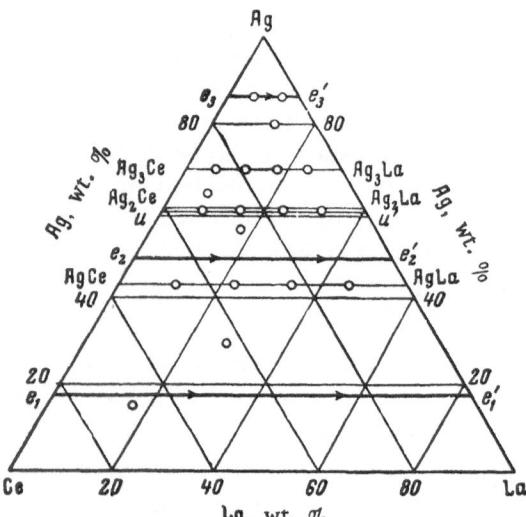

Fig. 52. Phase diagram of ternary system Ag—La—Ce with sections through binary compounds LaAg$_3$—CeAg$_3$ and LaAg—CeAg.

Literature data [56] on the continuity of solid solutions in the system Ni$_3$P—Cu$_3$P require further checking, since these two compounds are not isomorphous; Ni$_3$P has a tetragonal structure and Cu$_3$P a hexagonal. Further investigations would be of scientific and practical interest.

In addition to the materials considered above, we cite interesting data on the existence of continuous solid solutions in the systems LaAg$_3$—CeAg$_3$ and LaAg$_2$—CeAg$_2$. These systems of lanthanum and cerium argentides were studied by the authors of [126] during an investigation of the ternary system Ag—La—Ce. In this system La and Ce, being direct analogs, form continuous mutual solid solutions, whereas silver forms series of compounds with these analogs, containing identical mole fractions of the latter element: LaAg$_3$, LaAg$_2$, and LaAg; CeAg$_3$, CeAg$_2$, and CeAg. The first and second pairs of compounds crystallize with a real maximum; they are isomorphous. All this creates conditions for total mutual solubility within these pairs of compounds. According to [126], the systems LaAg$_3$—CeAg$_3$ and LaAg—CeAg actually show continuous series of solid solutions. This is evident from the two-dimensional isothermal diagram (Fig. 52), where straight lines connect the compositions of two continuous solid solutions in the systems of binary compounds LaAg$_3$—CeAg$_3$ and LaAg—CeAg.

Solid Solutions of Ternary and More Complex Metallides

Many ternary metallides are described in a number of books on metallic compounds [54, 55, 56, 65, 114]. In Table 16 above, data taken from the literature [65, 109, 110] were given on the compositions of ternary and more complex compounds for the particular case of compounds with semiconductor properties. Together with others, this class of compounds is very interesting from the viewpoint of the general rules governing their formation and practical application. An especially detailed consideration of such compounds from the chemical viewpoint is given in [105].

Analysis of the compositions of these compounds with the general formula ABC$_2$ shows that element A is a Group IB or IIB metal, whereas constituents B and C are elements of Groups IVB, VB, and VIB. General questions of total exchange in systems of such ternary compounds have not as yet been specially investigated by anyone. However, there are some experimental data on the formation of solid solutions between ternary compounds of similar type.

Proceeding from our theses on the conditions of formation of continuous solid solutions between metallides, one can predict the formation of such solutions in cases where:

1) the ternary compounds are isomorphous;

2) the compounds have the same type of chemical bonding;

3) the metals acting as element A in the ternary compounds are electropositive analogs or nearest neighbors with nearly identical metallochemical properties;

4) constituents B and C are elements of Groups IV—VI, which are interchangeable in the structures of the compounds;

5) the ternary compounds are thermodynamically stable at all temperatures up to their melting points.

As is evident from Table 16, most compounds have cubic structures (ZnS-type). Some of these compounds have NaCl-type structures. Compounds of this type have covalent bonding, with characteristic high electron mobility. Ionic bonding in them is practically absent except when they contain sulfur. As one of the most electronegative elements of the triad S—Se—Te in Group VI, the latter favors the formation of compounds with ionic bonding.

It may be asserted that when the structures are isomorphous, the type of chemical bonding is the same, and the ternary compounds contain atoms of the same kind (Cu or Ag) or atoms with similar metallochemical proper-ties (As and Sb, Sb and Bi, Te and Se), these compounds should form continuous mutual solid solutions. This can be shown in the case of some experimentally studied or hypothetical binary systems of ternary compounds. Studies of solid solutions formed between isomorphous ternary compounds in combination with binary ones would also be interesting in this respect.

Ternary compounds with NaCl-type structures — $AgSbSe_2$, $AgSbTe_2$, $AgBiSe_2$, $AgBiTe_2$ — and binary systems based on these compounds were investigated in [191, 192].

In [191] the authors synthesized crystals of the compounds and established their structure type and atomic arrangement. The compounds $AgSbSe_2$, $AgSbTe_2$, $AgBiSe_2$, and $AgBiTe_2$ have NaCl-type structures at high temperatures, and the Ag and Bi(Sb) atoms in them are disordered. $AgSbSe_2$ and $AgSbTe_2$ retain the NaCl structure at room temperature, whereas the Ag and Bi atoms in $AgBiSe_2$ and $AgBiTe_2$ disordered only above room temperature. In the case of $AgBiSe_2$ the disordered NaCl-type structure persists only down to 287°; below this temperature it goes over to a rhombohedral phase. At still lower temperatures the structure of this compound is even more complex. $AgBiTe_2$ also occurs in two modifications: The high-temperature β-phase (above 428.5°) has a NaCl-type structure, whereas the low-temperature phase has a trigonal one.

These above-mentioned peculiarities in structure type of ternary compounds with respect to temperature affect the character of interaction of these compounds in the corresponding binary systems, as is shown in [192]. The authors of this paper undertook to investigate the equilibrium diagrams of a quaternary system of ternary compounds: $AgSbSe_2$—$AgSbTe_2$—$AgBiSe_2$—$AgBiTe_2$. To investigate this system, however, they needed to know six binary systems of ternary compounds, as shown in Fig. 53. These binary systems are $AgSbSe_2$—$AgSbTe_2$, $AgSbTe_2$—$AgBiTe_2$, $AgBiTe_2$—$AgBiSe_2$, $AgBiSe_2$—$AgSbSe_2$, and two systems along the corresponding diagonals: $AgSbSe_2$—$AgBiTe_2$ and $AgBiSe_2$—$AgSbTe_2$.

These six binary systems were studies [192] mainly by thermal and structural analyses. The following was established by this investigation.

1. In the system $AgSbSe_2$—$AgSbTe_2$ (Fig. 54), which has two pairs of identical atoms (Ag and Ag, Sb and Sb) and two pairs of analogous atoms (Se and Te, and Se and Te), continuous solid solutions with disordered NaCl-type structures exist in the entire temperature interval from that of in-cipient crystallization to room temperature. The alloys crystallize in a very narrow temperature interval (see the liquidus and solidus curves in Fig. 54). The lattice parameter (see Fig. 54, bottom) varies almost linearly, with some positive deviation from Vegard's law. This deviation is possibly due to a change in the nature of chemical bonding in the system, resulting from ad-dition of the more metallic element Te in the form of the compound $AgSbTe_2$ to the less metallic selenium in $AgSbSe_2$.

2. The binary system $AgSbSe_2$—$AgBiSe_2$, which consists of three pairs of identical atoms (Ag and Se) and two analogous atoms (Sb and Bi), also exhibits crystallization in the form of continuous solid solutions (Fig. 55). On cooling, owing to transition of $AgBiSe_2$ from a cubic to a rhombohedral struc-ture (at 287°), a break in solubility occurs. When bismuth is replaced by anti-mony in the system of these ternary compounds, the high-temperature phase with NaCl-type structure is quickly stabilized, as is obvious from the drop in the curves of $\beta \to \alpha$-conversion temperature versus composition in Fig. 55. The lattice parameter for alloys of this system deviates very little from

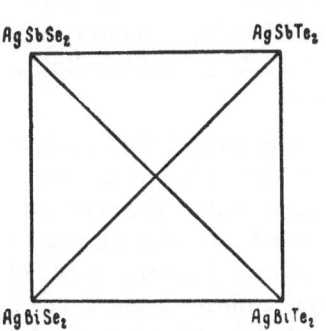

Fig. 53. Six binary systems of ternary compounds given at vertices of quadrangle for qua-ternary systems $AgSbSe_2$—$AgSbTe_2$—$AgBiSe_2$—$AgBiTe_2$.

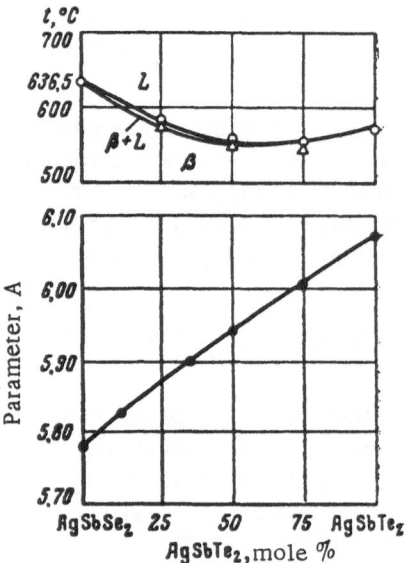

Fig. 54. Equilibrium and lattice parameter-composition diagrams of system $AgSbSe_2-AgSbTe_2$.

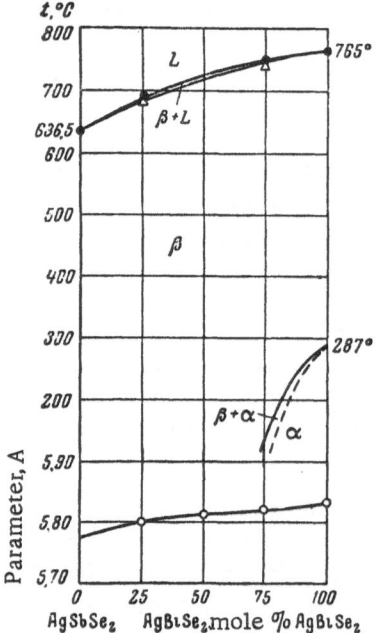

Fig. 55. Equilibrium diagram and curve of lattice parameter (of β-phase) versus composition for system $AgSbSe_2-AgBiSe_2$.

Vegard's law; some breaks in the line of the lattice parameter should exist for annealed alloys owing to transition of the cubic structure to rhombohedral in the region of alloys lying on the $AgBiSe_2$ side.

3. The three systems $AgSbTe_2-AgBiTe_2$, $AgSbSe_2-AgBiTe_2$, and $AgSbTe_2-AgBiSe_2$ are characterized by the fact that they contain the compounds $AgBiTe_2$ and $AgBiSe_2$, which each have two crystalline modifications, whereas the other compounds are isomorphous, and each of them has one modification with a NaCl-type structure. In conformity with this, all these three systems crystallize as continuous solid solutions; their fusibility diagrams are analogous to those of the above two systems, shown in Figs. 54 and 55. However, owing to the polymorphism of $AgBiTe_2$ and $AgBiSe_2$ in these three systems, $\beta \rightleftharpoons \alpha$-transformation occurs on the $AgBiTe_2$ (or $AgBiSe_2$) side, and the curves of transformation temperature versus composition are analogous to that shown in Fig. 55. Hence it is unnecessary to show the equilibrium diagrams of these systems here.

4. The sixth binary system, $AgBiSe_2-AgBiTe_2$, whose components each have two crystalline modifications, crystallizes as a continuous solid solution with a NaCl-type structure. In the solid state, however, solubility is discontinuous at all concentrations in this system. Its equilibrium diagram is shown in Fig. 56. It is evident from this that up to the temperatures of polymorphic transformation of the compounds ($AgBiSe_2$ and $AgBiTe_2$), the system corresponds to continuous β-solid solutions with a NaCl-type structure. Below these temperatures solubility is discontinuous, the two-phase $\alpha + \gamma$-region being formed by solid solutions with two types of structure (rhombohedral and trigonal). As is evident from the figure, however, this part of the diagram has not been studied in sufficient detail, and from it one can hardly judge the equilibrium character of phase distribution in the given system.

Based on analogous concepts, one may expect the formation of continuous solid solutions in other systems containing ternary compounds which meet the conditions of mutual solubility of metallides. Thus such solutions, composed of ternary compounds listed in Table 16, may be presumed present in systems having cubic structures (chalcopyrite):$CuInSe_2-CuInTe_2$, $AgInSe_2-AgInTe_2$, $ZnSiAs_2-ZnGeAs_2$, $CdSiP_2-CdGeP_2$, and other analogous systems.

One also can imagine such solid solutions formed between Laves, nickel arsenide, boride, carbide, and, finally, purely intermetallide phases with various types of structure. In a paper cited above [69] it was noted that 14 ternary compounds had been found among Laves phases. The authors hold that in such compounds element A is one of the Group II metals Mg, Ca, and Cd, whereas elements B and C are metals which can form electron compounds between themselves in such systems as Ni—Zn, Cu—Zn, Cu—Al, Ag—Zn, and Ag—Al. Such ternary compounds will form solid solutions when they are isomorphous, and atoms A, B, and C in the two compounds are of the same kind or similar in properties.

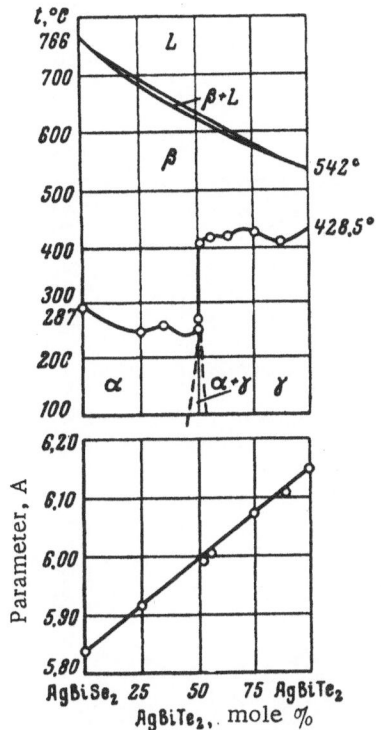

Fig. 56. Equilibrium diagram and curve of lattice parameter (of β-phase) versus composition for system $AgBiSe_2-AgBiTe_2$.

Among ternary carbide compounds, e.g., W_3Fe_3C, Mo_3Co_3C, Mo_3Mn_3C, and W_3Mn_3C [114], as well, if they are isomorphous, total mutual solubility obviously exists in the corresponding systems $W_3Fe_3C-Mo_3Co_3C$ and $Mo_3Mn_3C-W_3Mn_3C$. These systems have not been investigated experimentally; neither are there experimental data on the following systems of ternary carbides: $W_2Cr_{21}C_6-Mo_2Fe_{21}C_6$, $Mo_2Fe_{21}C_6-W_2Fe_{21}C_6$, and $W_2Cr_{21}C_6-W_2Fe_{21}C_6$. These three compounds are isomorphous and crystallize with $Cr_{23}C_6$-type structures. Hence they may be expected to display total mutual solubility. It is important to study these questions in order to individualize the phases in carbide analysis of steels and alloys.

The presence of continuous solid solutions may be presumed analogously in systems of the ternary borides Mo_2FeB_2, Mo_2CoB_2, and Mo_2NiB_2, which are isostructural compounds (U_3Si_2-type), and the borides Mo_2FeB_4, Mo_2CoB_4, and Mo_2NiB_4, which have the Cr_2NiB_4-type structures. However, these systems have not been studied experimentally, either.

In the paper cited above [170], the author analyzed questions of the formation of ternary silicides composed of two transition metals and silicon. Investigation of phase equilibria in the corresponding ternary systems of two transition metals (having unfilled 4d, 5d, and 6d electron levels) and silicon revealed the existence of 39 ternary silicides, whose structure types were determined.

Many of the ternary silicides given in the paper are isostructural and contain two transition metals which either are analogs or have similar metallochemical properties. Such are, for instance, the compounds MoFeSi, MoCoSi, MoNiSi, WFeSi, WCoSi, and WNiSi (MgZn-*type structures); $Cr_3Co_5Si_2$, $V_3Fe_5Si_2$, and $V_3Co_5Si_2$ (α-Mn-type structures); V_3Ni_2Si, Nb_3Ni_2Si, and Ta_3Ni_2Si (Ti_2Ni-type structures), etc. All these isomorphous compounds, when paired, should give continuous solid solutions, since in the first series of compounds Mo and W atoms, on the one hand, and Fe, Co, and Ni ones, on the other, are fully interchangeable in the structure of ternary silicides; Cr and V, and Co and Fe atoms in the second series of compounds, and V, Nb, and Ta atoms in the third series also are fully interchangeable. Hence the components of the systems MoFeSi–MoCoSi, MoFeSi–MoNiSi, MoFeSi–WFeSi, MoFeSi–WCoSi, MoFeSi–WNiSi, etc., should be regarded as mutually soluble, as in the systems of the last two series: $Cr_3Co_5Si_2-V_3Fe_5Si_2$, $Cr_3Co_5Si_2-V_3Co_5Si_2$, $V_3Fe_5Si_2-V_3Co_5Si_2$ and $V_3Ni_2Si-Nb_3Ni_2Si$, $V_3Ni_2Si-Ta_3Ni_2Si$, and $Nb_3Ni_2Si-Ta_2Ni_2Si$.

Further experimental investigations of these systems would be of interest and should confirm the hypotheses put forth above.

CHAPTER V

BINARY METALLIDE SYSTEMS WITH LIMITED SOLID SOLUTIONS

Conditions Limiting the Continuous Solubility of Metallides

As was noted above, if metallic compounds are of such physicochemical nature and structure that they do not satisfy the conditions of formation of continuous solid solutions, they form either limited solid solutions or mechanical mixtures devoid of solid solutions. In general, one may say that the mutual solubility of metallides in the solid state decreases with increasing difference in metallochemical properties of the constituent elements in these compounds. This corresponds in some measure to the degree of separation of these elements in the periodic system.

This general conclusion was reached in [123, 124] in the case of differing solubility of the metals Zn, Al, Sn, and Sb in the compound $MgCu_2$ and in [132] in the case of limited solubility of its components in the compound Ni_3Al. Of these metals Zn, which lies close to Mg and Cu in the periodic system, is the most soluble in $MgCu_2$; the solubility decreases from Zn to Al and Sn and is negligible for Sb — the most remote from magnesium and copper. The rules governing the variation in solubility of these metals in $MgCu_2$ are well illustrated by the solubility curves of Zn, Al, Sn, and Sb in $MgCu_2$, shown in Fig. 57 [124]. The authors of [123] explained these rules from the viewpoint of Brillouin-zone theory. The maximum content of metals in $MgCu_2$ corresponds to an electron concentration of 1.7 e/A, which in turn corresponds to filling of the first energy band in the $MgCu_2$ structure.

The same rules may be shown to apply in cases where the limited mutual solubility of compounds in metallide systems changes with increasing difference in metallochemical properties of their constituent elements. Discontinuities in solid solutions of binary compounds may appear also as a result of the presence in the corresponding systems of polymorphic modifications, intermediate phases, and especially ternary compounds. These important factors received attention in [114, 123, 137]. In some cases these circumstances upset the presumed quasi-binary relation between two compounds in ternary systems and preclude the formation of actual continuous solid solutions between compounds, even when the conditions for this, set forth above, are satisfied.

Examples of Metallide Systems with Limited Solubility

Below we shall consider some examples of experimental investigations dealing with limited solid solutions of metallides of various types and classes.

Here it is interesting to cite individual cases where limited solid solutions are formed between Kurnakov compounds. Such are the systems $FeCo-FePd_3$ and $FePd-FeCo$, formed from solid solutions of the ternary systems Fe−Co−Pd [193]. The components of this system are analogous metals which can give continuous solid solutions; however, there are discontinuities of solubility in metallide systems. This is due to the formation of nonisomorphous compounds and various modifications of iron and cobalt from solid solutions. The FeCo structure is body-centered, the $FePd_3$ face-centered, and the FePd tetragonal.

The discontinuity of solid solutions in this ternary system could be inferred from the superposition of regions of polymorphic transformation of the α- and γ-phases on the regions of formation of the compounds FePd, FeCo, and $FePd_3$, these phases having different types of structure in the studied sections. It was proved by experimental investigation in [193].

The corresponding equilibrium diagrams are shown in Fig. 58a for the section $FeCo-FePd_3$ and in Fig. 58b for the section FeCo−FePd. As is evident from the figure, both systems exhibit discontinuities in the mutual solubility of the compounds, and two- and three-phase regions. In some other cases one can show the presence of solubility gaps in systems of Kurnakov compounds, in which the limitation of solubility is due to the absence of structural isomorphism and the large difference in properties of the atoms.

85

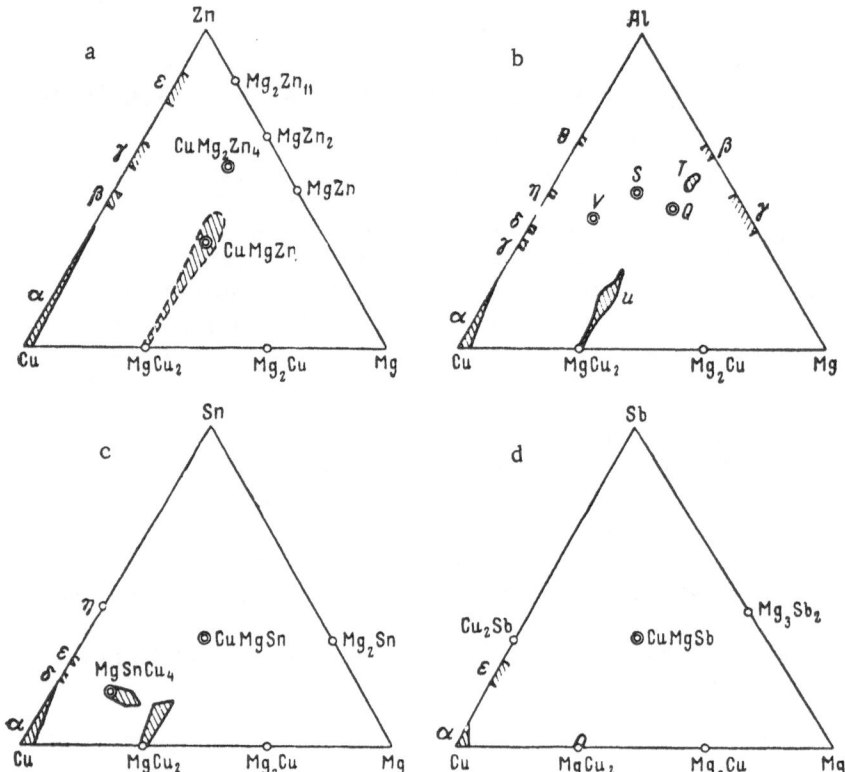

Fig. 57. Solubility curves of various metals in MgCu$_2$: a) zinc; b) aluminum;
c) tin; d) antimony.

Many systems of limited solid solutions can be formed between compounds crystallizing from the liquid phase. One striking example of such systems is the NbCr$_2$–NbNi$_3$ [194]. As is evident from the different atomic ratios of these compounds having niobium in common, the metals chromium and nickel are not analogs, so that these compounds also are nonisomorphous. For this reason, only limited solid solutions could be expected.

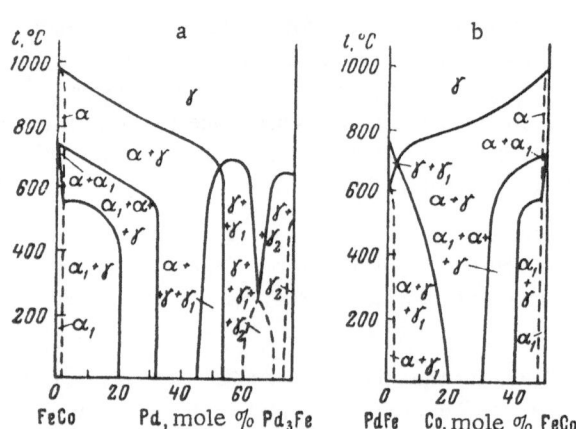

Fig. 58. Equilibrium diagrams of alloys in ternary system Fe–Co–Pd; a) section FeCo–FePd$_3$; b) section FeCo–FePd.

The equilibrium diagram of this system, plotted from data of [194], is shown in Fig. 59; here limited solubility and the formation of a eutectic mixture are observed. Since two polymorphic modifications of NbCr$_2$ are present (the high-temperature MgZn$_2$-type ε-form and the low-temperature MgCu$_2$-type β-form), an additional, intermediate two-phase β- + ε-region appears (Fig. 59). An analogous system with limited solubility of metallides is the ZrCr$_2$–ZrFe$_2$, studied by the authors of [153]. These two compounds are nonisomorphous, and hence discontinuities of solubility occur in this system. An example of a break in continuity due to the formation of a ternary compound in a metallide system is the section MgCu$_2$–MgZn$_2$, where, according to [195], the ternary intermediate phase Mg$_{2.5}$CuZn$_4$ is formed. For this reason the equilibrium diagram of this system appears in the form shown in Fig. 60.

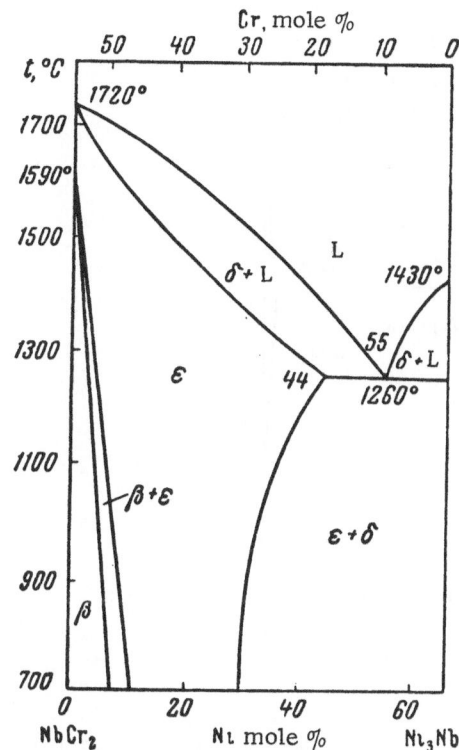

Fig. 59. Equilibrium diagram of system $NbCr_2-NbNi_3$.

A considerable difference in limited mutual solubility can be shown between a system in which the constituent elements are similar in structural and chemical properties and a system in which they are dissimilar. Such metallide systems, the UAl_2-ZrAl_2 and $UAl_2-U_3Si_2$ [196, 197], are shown in Fig. 61.

In the first system of compounds, where aluminum is the common element, the metals uranium and zirconium form continuous mutual solid solutions at high temperatures, but in the second system the compounds do not have the same atomic ratio; here uranium is the common element, whereas aluminum and silicon have different metallochemical properties. In this case it will be noted that UAl_2 and $ZrAl_2$ have the structure of Laves phases and are nonisomorphous; UAl_2 is $MgCu_2$-type, $ZrAl_2$ is $MgZn_2$-type [196], and U_3Si_2 has a tetragonal structure. Hence one could expect only limited solid solutions in both systems.

Experimental investigations of these nonisostructural systems [196, 197] (see Fig. 61) confirm the above. Comparison of their equilibrium diagrams shows that the systems UAl_2-ZrAl_2 and $UAl_2-U_3Al_2$ have different degrees of mutual solubility. In the first system there is a broad region of solid solutions owing to the fact that U and Zr form continuous mutual solid solutions at high temperatures, and the possible continuity is upset only by the difference in structure type between the component compounds. In the second system of nonisomorphous compounds, $UAl_2-U_3Si_2$, where U is the common element, the constituent Al and Si atoms differ in properties and have extremely limited mutual solubility in the binary system containing them, and the two compounds do not have identical atomic ratios. Possibly they differ also in type of chemical bonding. All this leads to the conclusion that the metallide system $UAl_2-U_3Si_2$ has extremely low mutual solubility, reminding one of mutual solubility in the binary system Al—Si.

The system UAl_2-UFe_2, which is described in a recent paper on the ternary system U—Al—Fe [198], is analogous to the two systems given above, based on uranium metallides. The authors show, in full accordance

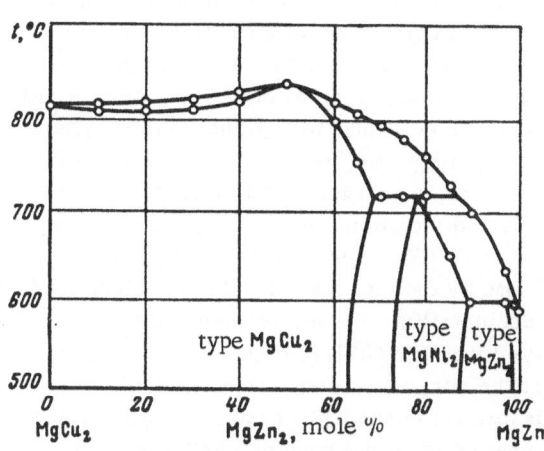

Fig. 60. Fusibility diagram of system $MgCu_2-MgZn_2$.

with theory, that the system UAl_2-UFe_2 is quasi-binary; owing to the presence of two metals, Al and Fe, with different metallochemical properties in these compounds, the mutual solubility of the metallides is limited. The large difference in electronegativity between Al and Fe is manifested in an additional reaction when these two metallides interact. In the UAl_2-UFe_2 system the ternary compound $U(FeAl)_2$ is formed; the latter has its own region of existence in the equilibrium diagram, which is plotted in Fig. 62 from the authors' data.

One example of the existence of limited metallide solid solutions is the system $Ni_3Al-NbNi_3$, investigated in [199, 200]. In this sytem the compound Ni_3Al is formed in a peritectic reaction and has a face-centered structure; $NbNi_3$ crystallizes with a real maximum and has a rhombic, β-$TiCu_3$-type structure. Thus, based both on structure and the content of atoms with different properties (Al and Nb) in this system, one could presume only limited mutual

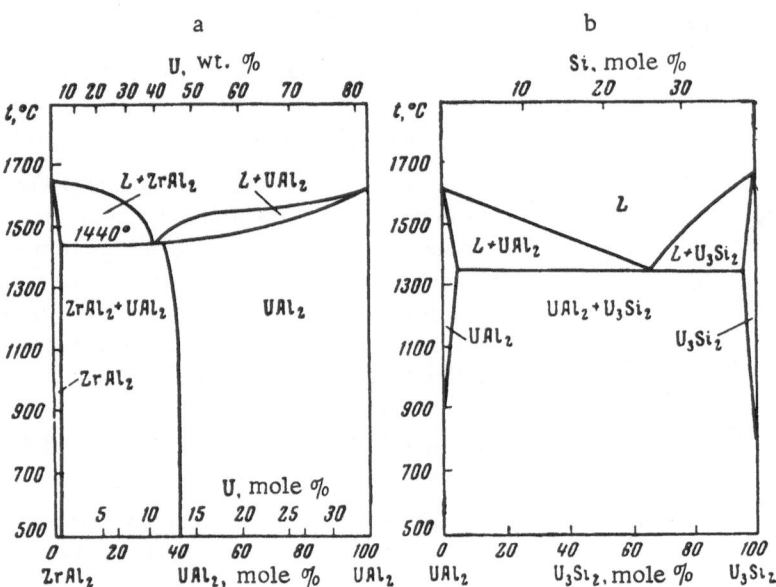

Fig. 61. Equilibrium diagrams of metallide systems: a) system UAl_2- $ZrAl_2$; b) system $UAl_2-U_3Si_2$.

solubility. Experimental investigation by thermal, microstructural, and x-ray structure analyses showed [199-200] that these two compounds give limited mutual solid solutions. $NbNi_3$ dissolves in Ni_3Al in amounts up to 40

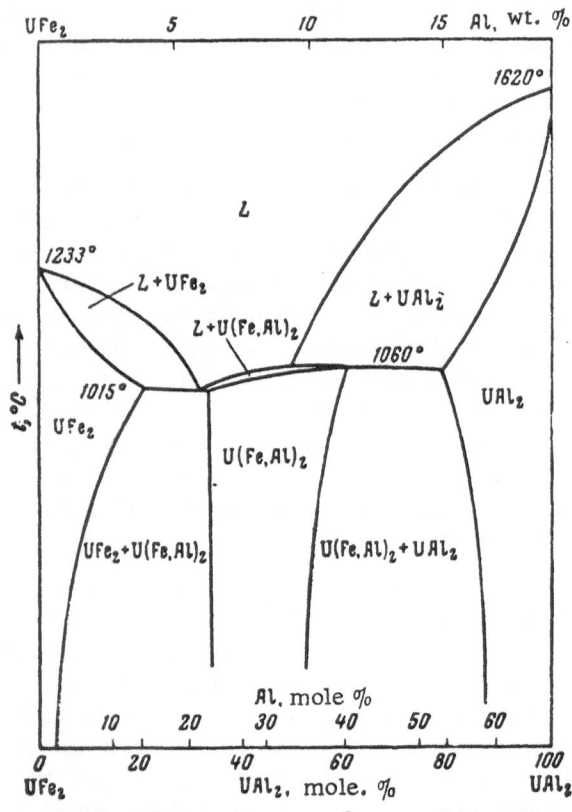

Fig. 62. Equilibrium diagram of system UAl_2-UFe_2.

wt. %, whereas Ni_3Al has quite negligible solubility in $NbNi_3$. Equilibrium and property—composition diagrams are shown in Fig. 63.

Many systems with limited solid solutions are encountered among borides, carbides, silicides, nitrides, etc.; many of them are interstial phases. These systems are partially described above [53-56, 78, 114].

Mutual solubility in such systems is limited when these compounds fail to satisfy one or more conditions of formation of continuous solid solutions between metallides. The determining factor in this respect is the absence of identical types of crystal structure and chemical bonding. Thus, for instance, in the system $CrSi_2-MoSi_2$, where the two disilicides are nonisomorphous ($CrSi_2$ is hexagonal and $MoSi_2$ tetragonal) and the Cr and Mo atoms are analogs, only limited solubility exists [55]. The limitation of solubility is favored also by the fact that $CrSi_2$ has covalent bonding and $MoSi_2$ metallic. The discontinuity of solubility in the system $FeSi_2-CrSi_2$ [201] is explained by the fact that the component compounds are nonisomorphous ($FeSi_2$ is tetragonal and $CrSi_2$ hexagonal) and have different types of chemical bonding. $FeSi_2$ has two modifications — α and β — of which the high-temperature β-form (above ~900°) has metallic bonding and

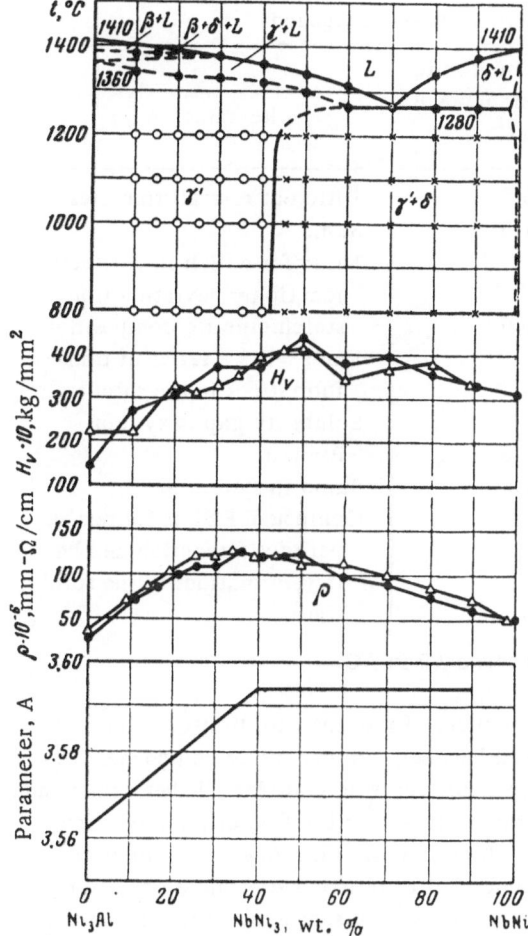

Fig. 63. Equilibrium and property-composition diagrams for system Ni₃Al—NbNi₃: ○ — one-phase structure; × — two-phase structure; △ — quenched state; ● — annealed state.

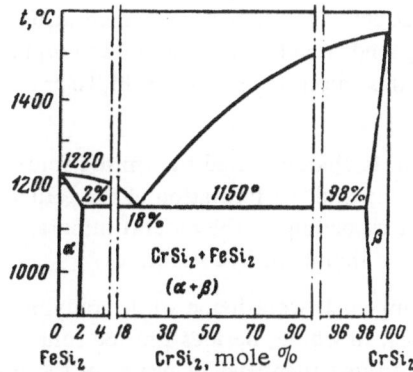

Fig. 64. Schematic equilibrium diagram of system FeSi₂—CrSi₂. FeSi₂ and CrSi₂— limited solid solutions α and β.

the low-temperature α-FeSi₂ (below the given temperature) covalent bonding.

This circumstance limits the mutual solubility of these Fe and Cr disilicides, as is shown in the diagram in Fig. 64, plotted from data of [201]. In this diagram no allowance is made for the two possible modifications of FeSi₂ mentioned above.

The same lack of continuity may be expected in the metallide system TiN—CeN, where the compounds have isomorphous structures (both b.c.c.), but the bonding is metallic in TiN and ionic in CeN. When interacting, the compounds display only limited mutual solubility. There are some systems with limited solid solutions, composed of superconductor compounds.

In a paper cited above [178] on solid solutions based on superconductor compounds, the authors studied several systems of such compounds and showed that they contain limited solid solutions. The latter are formed in systems where the compounds have different types of structure or contain atoms with markedly different metallochemical properties.

Systems of superconductor compounds which give limited mutual solid solutions [178] are listed in Table 31. As is evident, the enumerated systems display solubility gaps owing to the fact that the compounds are non-isostructural, and undergo complex transformations in the solid state. The equilibrium diagram of the system PdSb—PdBi, in which limited solid solutions occur, is shown in Fig. 65 as an illustration.

Among recent investigations of superconductor metallide systems with limited solid solutions, two papers dealing with the system Nb₃Sn—Nb₃Si [202, 203] are of interest.

Of these well-known compounds, Nb₃Sn has the β-W structure, whereas Nb₃Si [203] has an ordered, AuCu₃-type structure. Both compounds have superconductor properties. Since these two compounds are nonisomorphous, and moreover, Sn and Si are considerably different in their metallochemical properties, one might expect them to give only limited mutual solid solutions. When this system was investigated by determining the lattice parameter [202] and critical temperature T_C of transition to the superconducting state [203], the presence of such limited solubility was shown. As is evident from the lattice parameter values given in Fig. 66, the solubility limit of Nb₃Si in Nb₃Sn is about 50 mole %. Based on T_C data, although this quantity has not been determined accurately, the authors of [203] also find approximately 50 mole %. Up to this limit the critical temperature decreases smoothly as Nb₃Si is added to Nb₃Sn (see Fig. 66).

TABLE 31. Limited Mutual Solubility of Superconductor Compounds

Metallide system I — II	Solubility in compound*, mole %		Difference in atomic radius of elements B and C, %	Remarks
	I	II		
PtBi—PdBi	55	< 10	0.5	Different crystal structures
PdSb—PdBi	45	~ 8	11.5	Same
NiBi—PtBi	< 5	< 5	10.0	NiBi—formation in peritectic reaction — deviation from stoichiometric composition
NiBi—MnBi	< 5	< 5	4.5	MnBi — has a series of transformations in the solid state
PtBi—MnBi	< 5	< 5	5.8	Solubility gap in system Pt—Mn
PtBi—PtSn	< 10	~ 20	13.2	Same in system Bi—Sn
PtSb—CoSb	0	< 20	9.5	Compound PtSb is formed in peritectic reaction and has transformations in the solid state

*I — solubility of compound AC in AB; II — solubility of compound AB in AC.

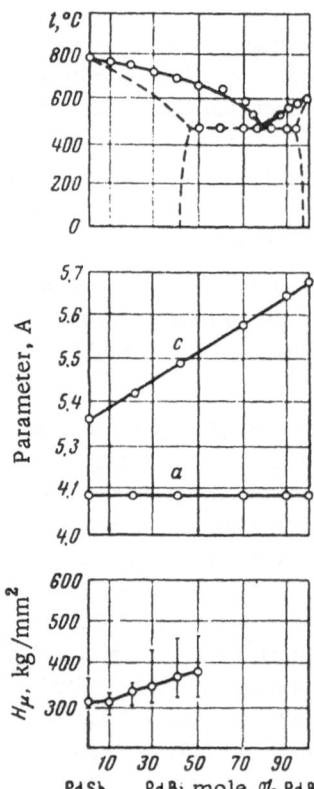

Fig. 65. Fusibility, parameter—composition, and microhardness—composition diagrams of system PdSb—PdBi.

A large number of systems with limited solid solutions are formed between semiconductor compounds [204-207]. It is impossible for us to consider all these cases, and we restrict ourselves here to a brief description of a few such systems. These systems are formed when their component compounds fail to satisfy one or more conditions of formation of continuous solid solutions. It was found that limited solid solutions are formed in the system Bi_2Se_3—Bi_2S_3, where the two compounds are nonisomorphous [181]. Two compounds, Bi_2Se_2S and Bi_2SeS_2, are formed in this system. Owing to the large difference in the atoms and their different stoichiometric ratios in the compounds, such systems as the $AlSb$—Al_2Te_3 [205], the $PbTe$—Bi_2Te and $SnTe$—Sb_2Te_3 [206], and the $PbSe$—Bi_2Se_3 [207] also contain only limited solid solutions. The latter are formed also when two binary compounds react to form a nonisomorphous ternary compound, as was shown in the system Bi_2Te_3—Bi_2S_3 [204].

The latter system is characterized by limited solubility of bismuth telluride and the formation of the ternary compound Bi_2Te_2S, corresponding to the natural mineral tetradymite, which has a rhombohedral structure.

Such ternary compounds were found while studying the system $PbSe$—Bi_2Se_3, in which, besides limited solubility, the formation of three independent phases of the following compositions was established: $3PbSe \cdot 2Bi_2Se_3$, $PbSe \cdot Bi_2Se_3$, and $PbSe \cdot 2Bi_2Se_3$ [207]. All these phases have semiconductor properties.

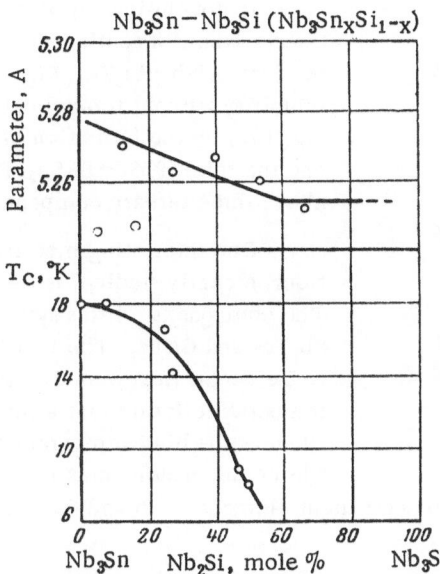

Fig. 66. Diagrams of critical temperature and lattice parameter versus composition for system Nb_3Sn-Nb_3Si with limited solid solutions.

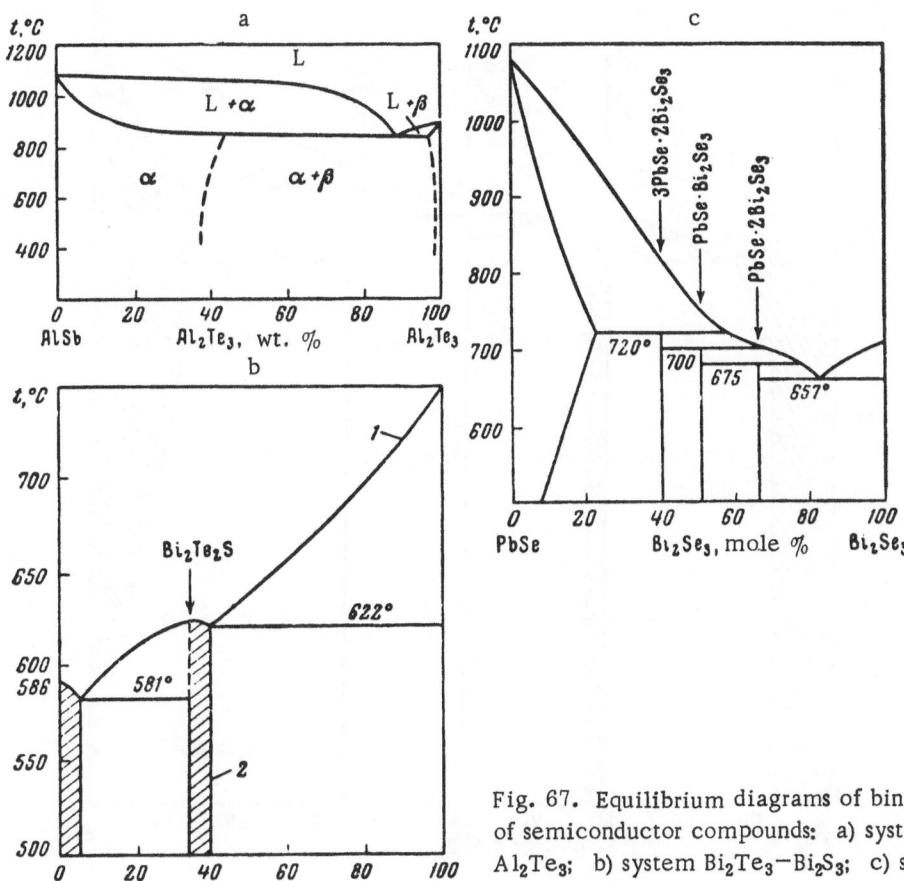

Fig. 67. Equilibrium diagrams of binary systems of semiconductor compounds: a) system $AlSb-Al_2Te_3$; b) system $Bi_2Te_3-Bi_2S_3$; c) system $PbSe-Bi_2Se_3$.

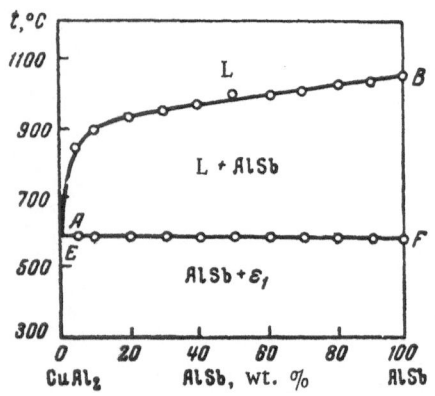

Fig. 68. Equilibrium diagram of system
CuAl₂−AlSb.

In conclusion, in order to illustrate the above, we show three systems of semiconductor compounds in Fig. 67, of which the first (AlSb−Al₂Te₃, Fig. 67a) is characterized by limited solubility without ternary compounds, the second (Bi₂Te₂−Bi₂S₃, Fig. 67b) by the formation of one ternary compound, Bi₂Te₂S, and the third (PbSe−Bi₂Se₃, Fig. 67c) by the formation of the above three ternary compounds.

One interesting metallide system with limited solid solutions, recently studied, is the Sb₂Te₃−GeTe, described in [282]. The components of this system are the semiconductor compounds Sb₂Te₃ and GeTe. The first has the tridymite structure, whereas GeTe occurs in two modifications, of which one — the low-temperature form — has a rhombohedral face-centered structure, whereas the high-temperature form has a NaCl-type structure. The common element is tellurium; the compounds have different atomic ratios, and the other two constituent elements − Sb and Ge − are quite dissimilar in metallo-chemical properties.

As was predictable, owing to nonisomorphism, the large difference in properties between Sb and Ge atoms, and the difference in atomic ratio of the compounds, one could not expect the formation of continuous

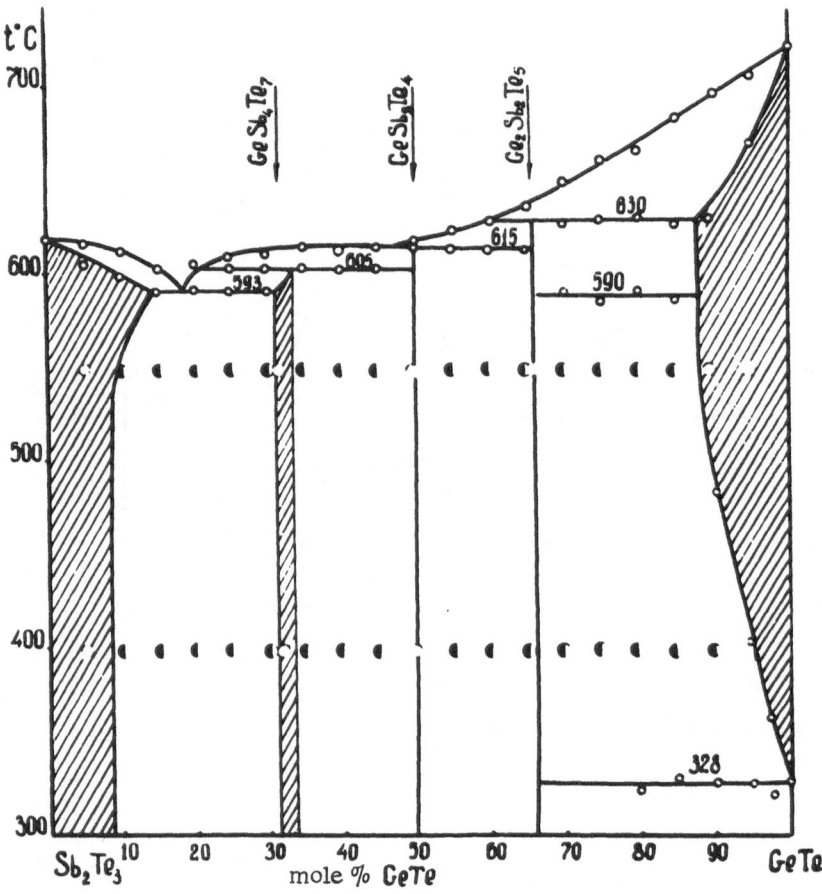

Fig. 69. Phase diagram of system Sb₂Te₃−GeTe.

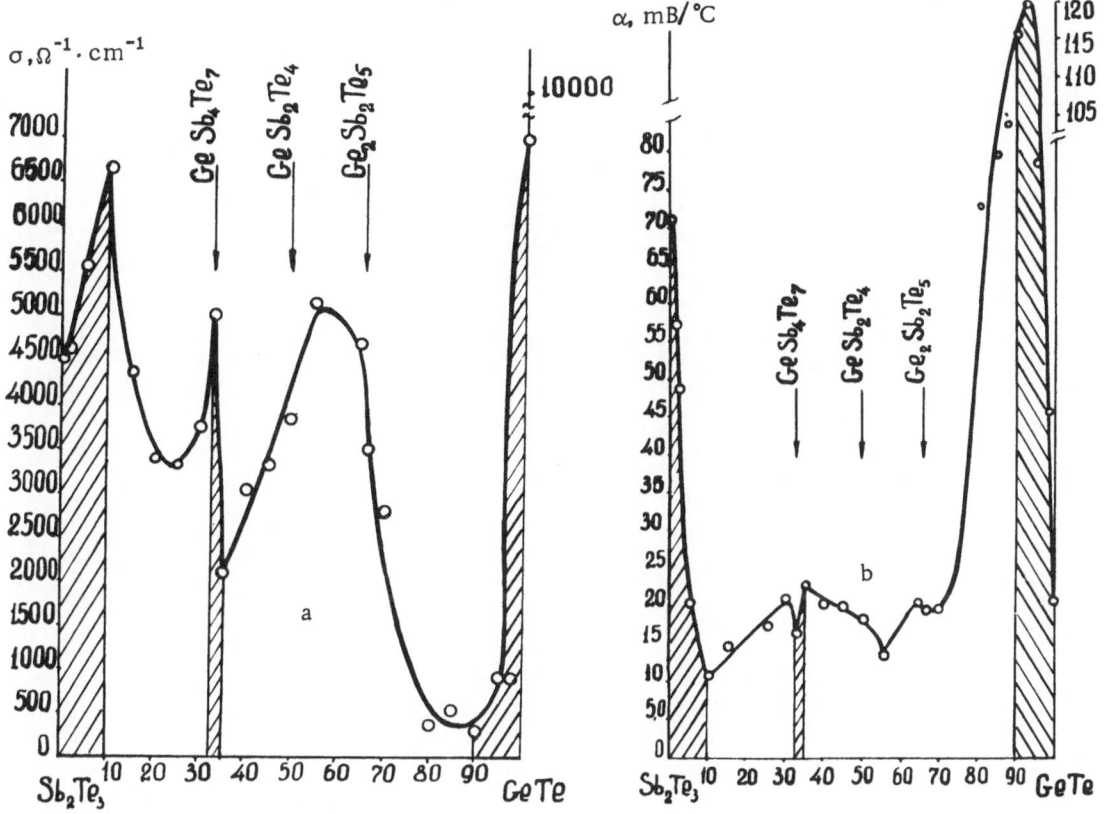

Fig. 70. a) Electrical conductivity of alloys of system Sb_2Te_3-GeTe; b) thermo-emf in system Sb_2Te_3-GeTe.

solid solutions in this system. Experimental study of the system Sb_2Te_3-GeTe [282] by thermal, microstructure, and x-ray analyses and physical-property measurements showed that it is characterized by limited mutual solubility. Moreover, the three compounds $GeSb_4Te_7$, $GeSb_2Te_4$, and $Ge_2Sb_2Te_5$ are formed in the system. The phase diagram of the system Sb_2Te_3-GeTe is shown in Fig. 69, from which it is evident that the system crystallizes in the form of limited mutual solid solutions, and the three compounds given above are formed in peritectic reactions.

Alloys of the system Sb_2Te_3-GeTe were studied with regard to electrical conductivity and thermo-emf, which are plotted against composition in Fig. 70ab.

As one may judge from these data, $GeSb_4Te_7$ shows a sharp electrical conductivity maximum and a shallow thermo-emf minimum. The other two compositions of compounds also exhibit distinctive properties in the property—composition diagrams.

Finally, compounds can be found which have different structures and whose constituent atoms differ markedly in metallochemical properties and lie far apart in the periodic system. In this case the total absence of mutual solubility may be presumed. An example of this is provided by the system $CuAl_2-AlSb$, shown in Fig. 68, which was studied in [208]. In this case the Cu, Al, and Sb atoms differ markedly among themselves, and the compounds are nonisomorphous; hence they exhibit total insolubility in the solid state.

CHAPTER VI

TERNARY AND MORE COMPLEX METALLIDE SYSTEMS

Very few experimental investigations of ternary and more complex systems composed of metallides have been conducted as yet. There are only a few papers dealing with these subjects. Here, as in the case of binary metallide systems, one can consider ternary solid solutions of both Kurnakov compounds formed from solid solutions and compounds crystallized from the liquid phase.

In [112, 113] many examples were given of possible ternary systems composed of Kurnakov compounds, in which continuous solid solutions should exist. Among them were noted the systems $FeNi_3 - CrNi_3 - MnNi_3$, $CuAu - CuPd - CuPt$, $CrFe - VFe - CrCo$, etc. No such experimentally studied systems are described in the literature.

One ternary system with limited solid solutions of Kurnakov compounds is the $VNi_3 - VCo_3 - VPd_3$ [139]. The given compounds are formed from γ-solid solutions of the corresponding binary systems.

It was noted above that the binary systems $VNi_3 - VCo_3$ and $VNi_3 - VPd_3$ form continuous solid solutions in this ternary system, whereas the system $VCo_3 - VPd_3$ exhibits limited solubility. Polythermal sections of the systems $VNi_3 - VCo_3$, $VCo_3 - VPd_3$, and $VNi_3 - VPd_3$ are shown in Fig. 71, and the equilibrium diagram of the ternary system $VNi_3 - VCo_3 - VPd_3$ at 600° is given in Fig. 72. The graphs show that in the system $VPd_3 - VCo_3$, formed from homogeneous γ-solid solutions, there is a solubility gap; this two-phase region $\gamma' + \gamma''$ lies in the part of the ternary system $VNi_3 - VCo_3 - VPd_3$, adjacent to the binary system of compounds VCo_3 and VPd_3.

The system $NbNi_3 - TaNi_3 - TiNi_3$ is an experimentally investigated ternary system in which continuous solid solutions are formed between compounds crystallizing from the liquid phase [209].

In an earlier study of an alloy composed of these three compounds [112, 113] it was shown that alloys of the entire system must be homogeneous metallide solid solutions. In a subsequent , more detailed investigation of the ternary system $NbNi_3 - TaNi_3 - TiNi_3$ [209] it was established that continuous ternary solid solutions actually exist in this system.

The above ternary system of three metallides is one of the triangular sections of the quaternary system $Ni - Nb - Ta - Ti$, obtained by sectioning a tetrahedral figure, as shown in Fig. 73. The triangular section produced in the tetrahedron in this way, having the three compounds $TiNi_3$, $NbNi_3$, and $TiNi_3$ located at the vertices, represents the ternary system $TiNi_3 - NbNi_3 - TaNi_3$, which we studied.

Alloys of this secondary ternary system, belonging to the sections shown in Fig. 74, were studied by thermal and structure analyses, as well as hardness measurements. Alloys of these sections crystallize as

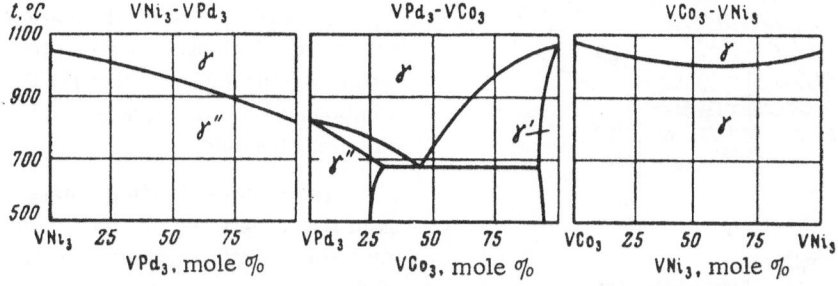

Fig. 71. Polythermal sections of systems $VNi_3 - VPd_3$, $VPd_3 - VCo_3$, and $VCo_3 - VNi_3$

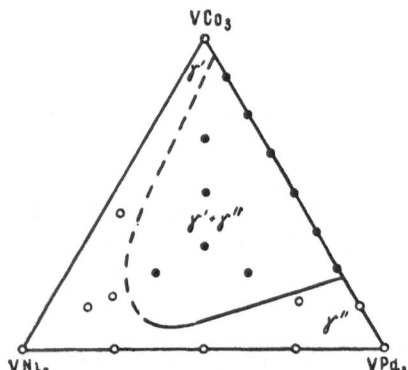

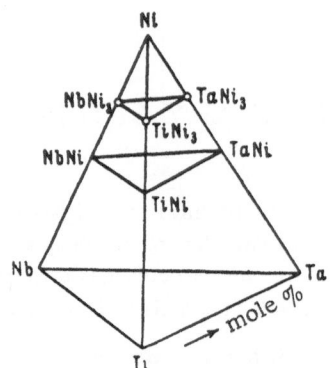

Fig. 72. Isothermal equilibrium diagram of ternary system $VNi_3-VCo_3-VPd_2$ at 600°: $\circ$ — homogeneous structure; $\bullet$ — two-phase structure; γ' — VCo_3 structure; γ'' — VPd_3 structure.

Fig. 73. Tetrahedration of quaternary system Ni—Nb—Ta—Ti with separation of two ternary metallide systems: $TiNi_3-NbNi_3-TaNi_3$ and TiNi—NbNi—TaNi.

continuous solid solutions, as the projections of the liquidus surface in the figure attest. In the annealed state all the alloys are homogeneous, and their hardness varies smoothly with composition. The possible existence of a slight gap caused by the hexagonal modification of $TiNi_3$ is indicated by the appearance of a two-phase region at the $TiNi_3$ vertex of the system $TiNi_3-NbNi_3-TaNi_3$.

In the quaternary system Ti—Nb—Ta—Ni under consideration, based on the presence of the compounds TiNi, NbNi, and TaNi in the corresponding binary systems [43], we propose that the second triangular section TiNi—NbNi—TaNi, as shown in Fig. 73, be investigated.

Analogously one may presume the formation of such solutions in the ternary system $TiNi_3-ZrNi_3-HfNi_3$. In this system, where the common metal is nickel, the metals titanium, zirconium, and hafnium are analogs in the three component compounds, and the compounds are isomorphous [43]. Based on this, one may consider that this ternary system is a continuous solid solution. This system has not been studied experimentally.

Continuous ternary solid solutions may be expected in the case of many other metallides: electron compounds, interstitial phases, superconductors, semiconductors, etc. For instance, such solutions may be presumed to exist in the systems $TiB_2-ZrB_2-HfB_2$, $TiB_2-VB_2-NbB_2$, TiC—ZrC—HfC, TiN—ZrN—HfN, and those of many other compounds containing elements with similar metallochemical properties. The two systems TiC—NbC—TaC and VC—NbC—TaC were studied in [210], whereas others are considered in a number of monographs [53-56, 78, 114].

An example of the formation of continuous ternary solid solutions among electron compounds is the system AgCd—AgZn—AgMg [123]. All these three compounds are isomorphous and have the same type of chemical bonding.

In the corresponding binary systems (see above), as in the ternary system AgCd—AgZn—AgMg, total mutual solubility is observed at temperatures above 500°.

This is attested by the results of investigating the lattice parameters of alloys in this ternary metal-

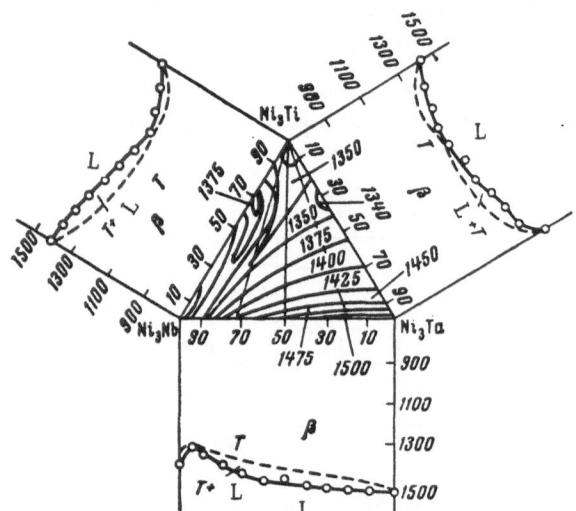

Fig. 74. Projection of liquidus surface from studied sections of ternary system $TiNi_3-NbNi_3-TaNi_3$.

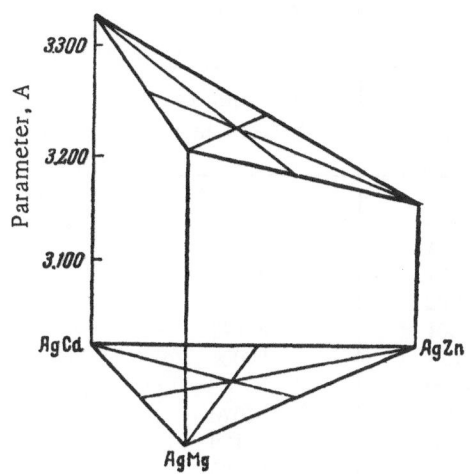

Fig. 75. Mutual solubility of β-electron compounds in system AgCd—AgZn—AgMg at 500°.

lide system; they are shown in Fig. 75. The lattice parameters of ternary solid solutions in the studied sections vary smoothly and continuously. Below 500° these electron compounds decompose, as is well known [43, and hence their ternary solid solutions also decompose.

Among ternary systems of superconductors compounds the system $Nb_3Sn—Ta_3Sn—V_3Sn$ [211] was studied experimentally [211]. All the metallide components of this system are superconductors; they are isomorphous and have β-W-type structures (see Table 13 above). In the corresponding binary systems they form continuous solid solutions. In view of the fact that the main conditions for formation of continuous solid solutions are met in this ternary system, one might expect total mutual solubility of all three compounds, Nb_3Sn, Ta_2Sn, and V_3Sn. In order to investigate the ternary system $Nb_3Sn—Ta_3Sn—V_3Sn$ [21], alloys were prepared from pure metal powders through mixing, pressing, and subsequent sintering in vacuo at 1200° for periods from 2 hr to 2 days.

X-ray structure investigations of samples prepared in this way showed the presence mainly of the same β-W structure in all the alloys. Some inhomogeneity in individual alloys was explained by insufficient homogenization on sintering. Curves of critical temperature of transition to the superconducting state versus composition also were taken. Both lattice parameters and critical temperatures in the ternary system vary smoothly, as is characteristic of continuous solid solutions.

Data on parameters and critical temperature for the studied alloys of the ternary system $Nb_3Sn—Ta_3Sn—V_3Sn$ are given in Tables 32 and 33.

A three-dimensional diagram of the variation in T_C for alloys of the ternary system $Nb_3Sn—Ta_3Sn—V_3Sn$, plotted from data of Table 31, is shown in Fig. 76. This diagram enables one to determine T_C values for any alloy composition of the system geometrically, by interpolation on the surface of the figure. The considerable importance of the limited experimental data on T_C for ternary metallides, obtained by the authors, consists in this.

Many ternary systems with limited solid solutions can exist among carbides, silicides, borides, and nitrides. Such systems containing hafnium carbides were investigated in [212]. The authors, using mainly x-ray structure analysis, studied six ternary carbide systems: HfC—TiC—VC, HfC—NbC—VC, HfC—TaC—VC, HfC—UC—ZrC, HfC—UC—NbC, and HfC—UC—TaC. All these carbides have face-centered structures (NaCl-type) and are metallic compounds.

In full accord with our above thesis, a break in continuity may occur in a ternary metallide system of isomorphous compounds having the same type of chemical bonding if their constituent elements differ in metallochemical properties. Of the systems given above, the constituents Hf and V, as well as Hf and U, differ in their properties and position in the periodic system. They form only limited mutual solid solutions. In the six ternary systems listed above, therefore, only limited solid solutions exist owing to limited solubility in the binary systems HfC—VC and HfC—UC.

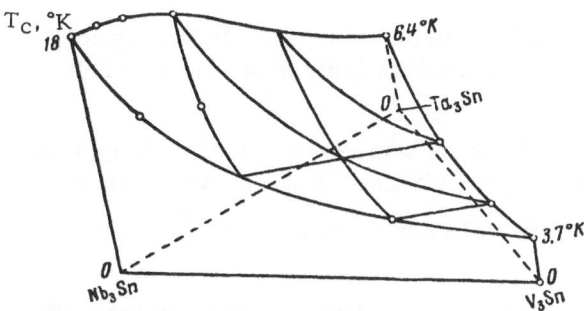

Fig. 76. Three-dimensional diagram of critical temperature T_C for alloys of ternary metallide system $Nb_3Sn—Ta_3Sn—V_3Sn$.

TABLE 32. Lattice Parameters of Ternary Solid Solutions of the Metallide System
$Nb_3Sn-Ta_3Sn-V_3Sn$

Composition of compounds in solid solutions	Parameter, A		According to Vegard's law
	according to literature data	experiment	
Nb_3Sn	5.289	5.289±001	—
Nb_2TaSn	—	5.287±001	5.285
$NbTa_2Sn$	—	5.280±001	5.282
Ta_3Sn	5.276	5.278±002	—
Ta_2VSn	—	5.174±001	5.172
TaV_2Sn	—	5.041±001	5.006
V_3Sn	4.94	4.96 ±002	—
V_2NbSn	—	5.115±004	5.070
VNb_2Sn	—	5.171±003	5.179
$NbVTaSn$	—	5.175±001	5.176

TABLE 33. Critical Temperatures of Transition to the Superconducting State of Solid
Solutions of the System $Nb_3Sn-Ta_3Sn-V_3Sn$

Composition of compounds in solid solutions	Critical temperature, °K		Composition of compounds in solid solutions	Critical temperature, °K	
	experiment	calculation		experiment	calculation
Nb_3Sn	18.0	31.0	TaV_2Sn	2.8	3.2
$Nb_{2.75}Ta_{.25}Sn$	17.8	—	V_3Sn	3.8	3.8
$Nb_{2.5}Ta_{.5}Sn$	17.6	24.8	V_2NbSn	5.5	7.0
Nb_2TaSn	16.4	16.5	VNb_2Sn	9.8	14.3
$NbTa_2Sn$	10.8	9.9	$V_{.5}Nb_{2.5}Sn$	14.2	19.9
Ta_3Sn	6.4	6.2	$NbVTaSn$	6.2	—
Ta_2VSn	3.7	3.8	$Nb_2Ta_{.5}V_{.5}Sn$	12.2	14.5

Two ternary equilibrium diagrams of the carbide systems HfC−NbC−VC and HfC−ZrC−UC, plotted from data of [212], are given in Fig. 77. They are shown for the 2050° isotherm. Curves delimiting the one- and two-phase regions and lattice isoparameters are given in the diagrams. The other four HfC-containing systems given above are constructed analogously.

The presence of limited carbide solid solutions in the ternary system UC−ThC−ZrC also was reported [213]. In this system there is a break in continuity owing to limited mutual solubility in the binary carbide system ThC−ZrC.

Another example of limited solid solutions based on metallides is the ternary system $CrNi_3-TiNi_3-Ni_3Al$ [214]. One of these compounds ($CrNi_3$) is formed from solid solutions in $TiNi_3$ and Ni_3Al on crystallization.

The phase diagrams of this system of 1000 and 750° were established by a detailed investigation using x-ray and microstructure analyses.

The 750° isotherm, plotted from data of [214], is shown in Fig. 78; it is evident from this that there are considerable regions of limited solid solutions ($\gamma' + \gamma$) based on the compounds Ni_3Al and $CrNi_3$, whereas these components are nearly insoluble in the compound $TiNi_3$. The existence of limited solid solutions in these compounds is consistent with the general rules of formation of solid solutions among metallides.

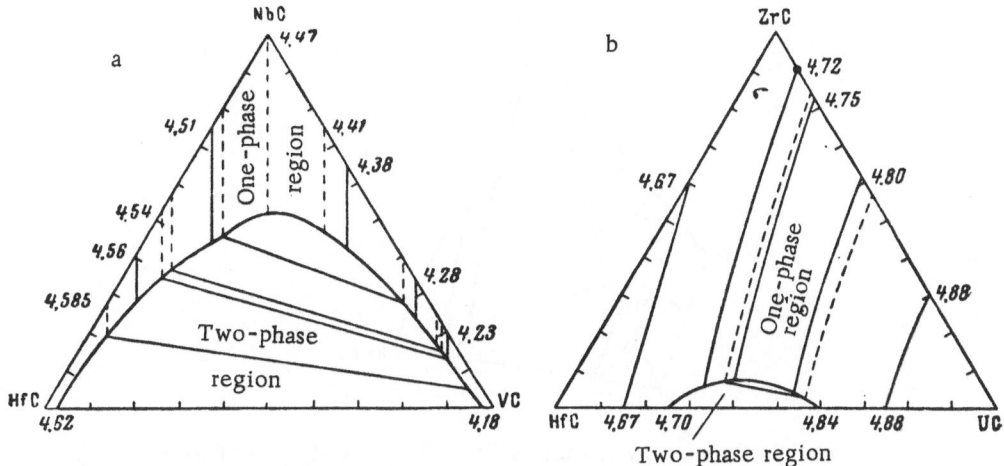

Fig. 77. Limited ternary solid solutions at 2050°. a) Monocarbides NbC−VC−HfC; b) monocarbides ZrC−HfC−UC. Numbers on sides are lattice parameters in A: —— experiment; - - - - calculation.

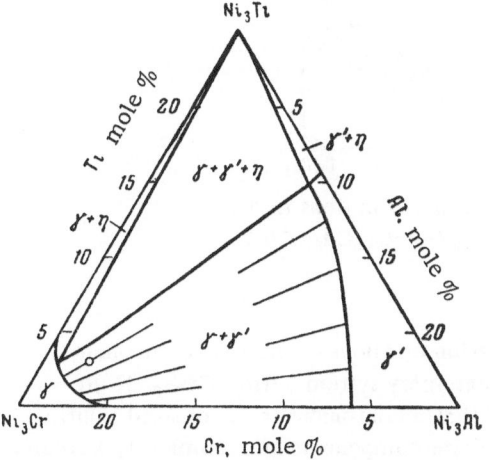

Fig. 78. Phase diagram of ternary system CrNi₃−TiNi₃−Ni₃Al at 750°.

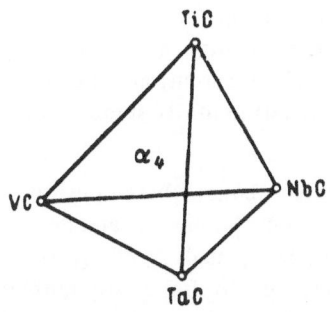

Fig. 79. Continuous quaternary solid solutions (α_4) of monocarbides TiC−VC−NbC−TaC.

Further investigation of equilibria among metallides in systems more complex than ternary may show that solid solutions based on compounds can exist in quaternary and more complex metallide systems. After applying the main conditions determining formation of solid solutions among metallides, one can predict the formation of continuous or limited solid solutions in such systems without experimental data. In this case the components may be either compounds with purely metallic bonding, interstitial phases, or compounds with another type of chemical bonding.

A tetrahedral diagram of the quaternary monocarbide system TiC−VC−NbC−TaC is shown in Fig. 79. All these carbides are isomorphous and have NaCl-type structures. In these compounds the common atom is carbon; the metals Ti, V, Nb, and Ta, nearly identical in metallochemical properties, form continuous mutual solid solutions [43]. Based on this, the entire system will correspond to homogeneous quaternary solid solutions (α_4), as shown in Fig. 79. No one has studied such a quaternary system as yet, although investigation of alloys of this system would be interesting from the viewpoint of establishing the rules governing the variation of properties in complex systems of interstitial phases. The same hypotheses regarding the formation of complex solid solutions among many-component systems of titanium, vanadium, niobium, tantalum, zirconium, hafnium, and uranium monocarbides were advanced in [215]. To an equal degree one may assert that such solutions are possible in quaternary systems of borides, silicides, nitrides, etc.

An example of quaternary metallide systems with limited solid solutions is the system TiNi₃−CrNi₃−FeNi₃−Ni₃Al.

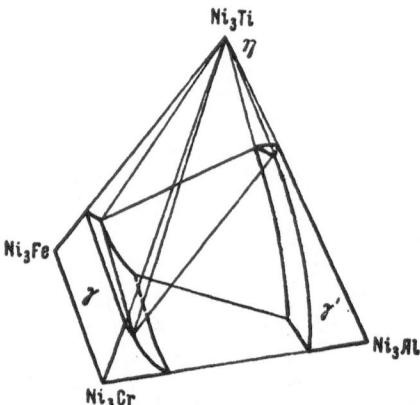

Fig. 80. Quaternary systems CrNi₃—TiNi₃—FeNi₃—Ni₃Al with limited solid solutions

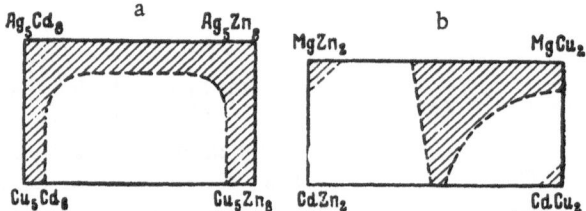

Fig. 81. Diagram of formation of limited solid solutions in quaternary metallide systems: a) $Ag_5Cd_8-Ag_5Zn_8-Cu_5Cd_8-Cu_5Zn_8$; b) $MgZn_2-MgCu_2-CdZn_2-Cu_2Cd$.

It was studied experimentally in [216] in connection with an investigation of alloys of the five-component system Ni—Cr—Fe—Ti—Al. The equilibrium diagram of the quasi-quaternary system $TiNi_3-CrNi_3-Ni_3Fe-Ni_3Al$, plotted from data of [213], is shown in Fig. 80; it is evident from this that a continuous series of solid solutions exists between $CrNi_3$ and $FeNi_3$, Ni_3Al has limited solubility, and all these components are completely insoluble in $TiNi_3$. One also should point out the work of [123], where the authors studied the two quaternary metallide systems $Ag_5Cd_8-Ag_5Zn_8-Cu_5Cd_8-Cu_5Zn$ and $MgZn_2-MgCu_2-CdZn_2-CdCu_2$, which form limited solutions. The phase diagrams of these systems are shown in Fig. 81.

The formation of solid solutions (continuous or limited) is not restricted to many-component systems consisting of binary metallic compounds. One may propose that such solutions can be formed in many-component systems consisting of ternary and more complex compounds. If the ternary compounds are isomorphous and composed of analogous elements which conform to the rules of formation of solid solutions, they should, despite their complexity, give mutual solid solutions. A case in point is the experimentally studied quaternary system $AgSbSe_2-AgSbTe_2-AgBiSe_2-AgBiTe_2$ [192].

As is evident, the system consists of the four ternary compounds considered above, whose constituent elements include one common metal — silver — and the pairs of analogous elements Sb and Bi, and Se and Te; the latter form continuous mutual solid solutions [43]. According to [192], the compounds $AgBiSe_2$, $AgBiTe_2$, $AgSbSe_2$, and $AgSbTe_2$ have cubic structures (NaCl-type) at high temperatures; $AgBiSe_2$ has a polymorphic rhombohedral modification at temperatures below 287°, whereas at 428.5° $AgBiTe_2$ also goes over to another (trigonal) modification. As was noted above, a preliminary investigation of the six quasi-binary subsystems of

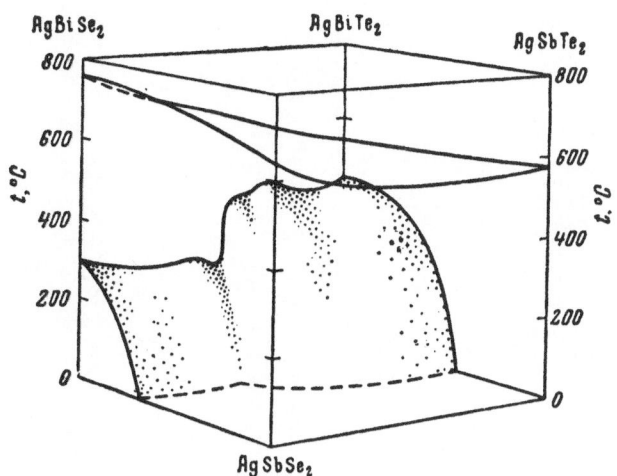

Fig. 82. Three-dimensional diagram of quaternary metallide system $AgSbSe_2-AgSbTe_2-AgBiSe_2-AgBiTe_2$.

ternary compounds: $AgSbSe_2-AgSbTe_2$, $AgSbSe_2-AgBiSe_2$, $AgSbTe_2-AgBiTe_2$, $AgSbSe_2-AgBiTe_2$, $AgSbTe_2-AgBiSe_2$, and $AgBiSe_2-AgBiTe_2$, showed [192] that continuous solid solutions exist in all six at high temperatures. In the corresponding systems containing $AgBiSe_2$ and $AgBiTe_2$, a solubility gap occurs on account of polymorphic modifications of $AgBiSe_2$ and $AgBiTe_2$.

The authors plotted a three-dimensional diagram (Fig. 82) of the quaternary system of these ternary compounds by generalizing the data on the studied quasi-binary systems. As is evident, continuous quaternary solid solutions of compounds exist in the system $AgSbSe_2-AgSbTe_2-AgBiSe_2-AgBiTe_2$ at high temperatures, whereas a break in continuity is observed at low temperatures on account of polymorphic modifications of $AgBiSe_2$ and $AgBiTe_2$.

Thus investigations of equilibria in simple and many-component systems of metallides show that they can react among themselves. Continuous and limited metallide solid solutions and simple mechanical mixtures are formed in the systems by such reactions.

CHAPTER VII

SOME PROPERTIES OF METALLIDES AND METALLIDE SYSTEMS

Characteristics of Metallides in Property — Composition Diagrams of Metallic Systems

Metallic compounds, as noted above, are in many respects individual substances with definite atomic ratios. The properties of such compounds differ from those of their components.

This thesis, set forth by N. S. Kurnakov in the beginning of this century, is fundamental for determining the compositions of metallides formed in metallic systems. The indisputable fact or, as one may say, generally accepted Kurnakov law is that these compounds of constant composition — daltonides — are represented by singular points of maxima and minima in property—chemical composition diagrams. As regards compounds of variable composition — berthollides — their properties vary smoothly in such diagrams, and the regions of existence of phases of variable composition are characterized by breaks, or changes in direction of the property — composition curves.

When considering actual property — composition diagrams, one should note that we may observe minima at the singular points for a given property of compounds in some system and maxima for other properties. This follows from the individual properties of the compounds, which are connected with their crystal structure and character of chemical bonding. It has been established that compounds with metallic bonding (intermetallides), which can, as a rule, form solid solutions with their components, show minima in diagrams of hardness and electrical resistivity versus composition, whereas compounds with covalent bonding, which usually do not form solid solutions with their components, have maxima of hardness and electrical resistivity in such diagrams.

The above is illustrated in Fig. 83 by two diagrams. One of them is the electrical resistivity — composition diagram (Fig. 83a) of the system Fe—V, where the minimum of this property corresponds to the compound VFe_3, formed from solid solutions. The same diagram for the system Mg—Sn is shown in Fig. 83b; it has an electrical resistivity maximum for the compound Mg_2Sn. Of these compounds, the first is a Kurnakov compound with metallic bonding and is formed from solid solution, whereas the second, a typical semiconductor compound with covalent bonding, is formed on crystallization. This characteristic difference in singular points corresponding to the compositions of compounds with different types of chemical bonding has not received special attention until now. In courses on metallurgy and physicochemical analysis, diagrams of hardness and electrical resistivity versus composition, showing minima for intermetallides, are the only property—composition diagrams given in the usual exposition of Kurnakov's laws; the properties of other metallides are excluded from the field of view, particularly semiconductor compounds, which show maxima on the property—composition diagrams (Fig. 83).

When considering the properties of individual compounds and equilibrium systems based on them, one must pay special attention to practically important properties, which determine their possible applications. These include mechanical, physical, chemical, and other properties. Many metallides have high hardness, heat resistance, resistance to oxidation on heating to high temperatures, and high corrosion resistance in aggressive media. A number of compounds have superconductor, semicondutor, magnetic, and other special physical properties.

It is not possible to analyze all the properties of such compounds in this monograph. They are partly described above in connection with the compositions of the compounds and are given in various monographs [77, 78, 217, 220] and handbooks [219]. Here we limit ourselves to certain characteristics of practical importance in separate groups of compounds.

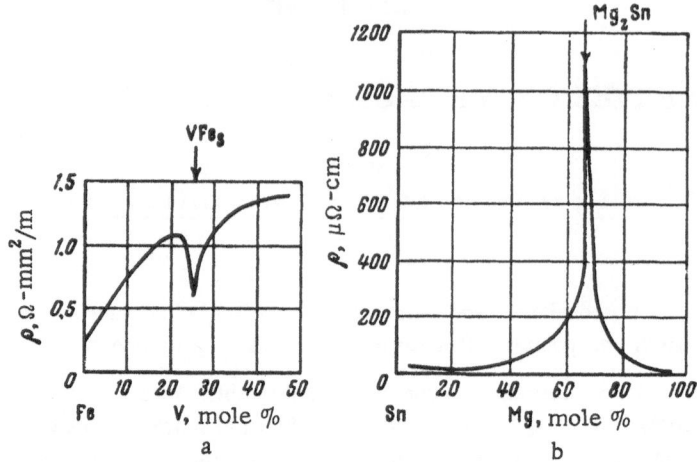

Fig. 83. Diagram of electrical resistivity ρ versus composition;
a) system V—Fe containing Kurnakov compound VFe_3; b) system
Mg—Sn containing semiconductor compound Mg_2Sn.

Some Properties of Heat-Resistant Intermetallides

These properties of metallides are considered most fully and comprehensively in reports to a symposium on the mechanical properties of metallic compounds[218]. Reports by Westbrook, Schwab, Savitskii, the author of the present book, et al., dealing with various mechanical, physical, and technological properties of metallic compounds, were published in the transactions of this symposium.

In the beginning of this century, Kurnakov and his co-workers [14, 15] found that metallides, as a rule, are harder than their components. Moreover, these compounds are exceptionally brittle at low temperatures. At one time it was supposed that all metallides were brittle at room temperature. Recent investigations of these properties in many compounds, however, led to the discovery of some compounds which are quite plastic when cold. Thus, for instance, it was found that of all intermediate phases in the nickel—aluminum system, the compound Ni_3Al is the most plastic, or deformable when cold [218]. This compound is the most heat-resistant material in the system Ni—Al [221].

It was shown in [221] that the compound TiNi, contrary to other metallides in this system, also has plastic properties at room temperature. This is evident from the following data on the mechanical properties of TiNi at 20°: σ_b = 85-95 kg/mm^2; δ = 15%; a_n = 3.8 $kg\text{-}m/cm^2$. At 649° this compound retains a tensile strength of the order of 80-85 kg/mm^2.

Other metallide compounds with plastic properties, deformable at room temperature, have been reported. Thus, for instance, it is stated in [218] that $ZrNi_4$ can be worked; in [223] it is said that some intermediate phases in systems of niobium and tantalum with platinum-group metals, although very hard, are sufficiently plastic to be worked. Such phases in these systems are those based on the compounds NbRh and NbIr, and TaRh and TaIr. The authors of the above studies, being unable to explain these very interesting phenomena in detail for the present, can only state them. Without doubt some metallides are plastic at low temperatures. The decisive role in this respect apparently is played by the character of chemical bonding, the type of crystal structure, and the degree of order of atoms in the crystals. One may say that compounds formed by components with nearly identical metallochemical properties, metallic bonding, and ordered arrangements of atoms are more plastic than those formed by elements with very different metallochemical properties and complex crystal structures. Among the relatively plastic metallides, one should distinguish certain Kurnakov compounds, formed in the solid state, the most brittle compounds, which contain interstitial atoms, and compounds with covalent, mixed covalent-metallic, or covalent-ionic bonding.

In cases of covalent or intermediate types of bonding, the interatomic bonds are more strongly directed than in pure intermetallides. The "cationic" and "anionic" bonds between electropositive and electronegative atoms in covalent compounds and the varied degree of covalent and ionic bonding in them determine their high hardness. Their high brittleness also follows from this.

However, all brittle metallic compounds have one special property: On heating above certain temperatures they become plastic, and can undergo plastic deformation while hot. In a paper published in 1923 [224], certain rules were established governing the variation in hardness and coefficients of contraction and impact with test temperature. It was shown that metallic compounds have well-marked transition temperatures, at which they go over from the brittle to the plastic state.

In one group of compounds this transition is abrupt; in another group it is more gradual; for some compounds the critical temperature of transition from the brittle to the plastic state is equal to 0.70 of the absolute melting point (T_m), whereas for others it is $0.96 T_m$. It was found that this critical temperature is about $0.88 T_m$ on the average.

These ideas regarding critical temperatures of transition from the brittle to the plastic state were developed by subsequent studies, and considerable experimental material accumulated along these lines. These questions are considered in generalized form in [217, 220].

Owing to the increased strength of chemical bonding in metallides as compared with their components, their hardness, as well as other mechanical properties, is increased. One sensitive index of change in strength of interatomic bonds is the modulus of normal elasticity. It was shown [218] that compounds with metallic bonding have smaller moduli of elasticity (e.g., CuZn and AuCu) than compounds with covalent bonding (e.g., Cu_9Al_4, GaAs, etc.). Many metallides have exceptionally high heat resistance [220-221].

Some mechanical properties at room temperature and higher are given in Table 34 for those intermetallides which are important as prospective heat-resistant materials. Data on structure type and melting point also are given in the table.

This table could be extended by including data on the mechanical properties of many refractory metallides. However, such data have already been set forth in detail in [217-220].

The mechanical and other properties of a relatively new class of compounds, known as beryllides [225], are of some interest. These compounds, formed between electropositive transition metals and electronegative beryllium, have unusually high melting points and oxidation resistance on heating to 1200-1600°. The strength of beryllides also is high, up to these temperatures, and they are very interesting from this viewpoint.

These metallides are characterized by an exceptionally high ratio of beryllium to transition metal atoms in their compositions. Thus many beryllides have such general formulas as Me_2Be_{17}, $MeBe_{11}$, $MeBe_{12}$, and $MeBe_{13}$, which sets them apart from most other metallides. Titanium, zirconium, niobium, tantalum, and molybdenum beryllides having atomic ratios corresponding approximately to these formulas were studied in [225]. The compositions of such compounds are given in Table 35, together with some of their properties — density of sintered samples, melting point, working test temperature, and also tensile strength (by bending) and modulus of elasticity at the latter. As is evident from these data, the compounds studied have melting points of 1700° or more; many of them have quite high tensile strength (from 4.1 to 42 kg/mm^2) at 1250-1500°. There are indications that zirconium and niobium beryllides react with moisture.

Compounds of electronegative aluminum with electropositive transition metals, which we call aluminides, are similarly interesting. This group of compounds, similarly to beryllides, includes a large number which are enriched in aluminum atoms, i.e., the ratio of aluminum atoms to those of other metals is larger than 1:1. In these compounds, which do not by any means correspond to normal valence relations, electrons of the aluminum atoms occupy the unfilled d orbitals of the transition metal atoms to some degree, and certain types of complex crystal structure are formed as a result. Many aluminides have special mechanical and physical properties; they include the heat-resistant and high-temperature oxidation-resistant compounds Ni_3Al, $NiAl$, $CoAl$, Ti_3Al,

TABLE 34. Some Properties of Intermetallides Which are Promising as Heat-Resistant Materials

Соединение	Structure type	M.p., $^\circ$C	Mechanical properties at		Remarks
			room temperature	elevated temperatures	
Ni_3Al	Cu_3Au	1385 *	$\sigma_b = 28.1$ kg/mm^2; $\delta = 1\%$; work of impact 11.5–18.5 kg/cm	$\sigma_b = 14.05$ at 815°; σ_{100} at 815° = 6.05 kg/mm^2	Grain size and deviation from stoichiometric composition affect properties appreciably
Ti_3Al	Hex.	Formed from solid solution, ∼1200	—	—	Worked in hot state (becomes malleable at t = 1100–1200°)
NiAl	CsCl	1650	M. r. = 101.0 kg/mm^2	M. r. = 49.2 kg/mm^2 at 950°	σ_S found at ∼950°; alloy malleable at 1300°. Grain size stable on annealing at 1000°
NiAl	CsCl	1650	M. r. = 91.5 kg/mm^2	M.r. = 21.2 kg/mm^2 at 1050°	Melted with additions of Mo, TiC, MoAl$_2$
NiAl	CsCl	1650	—	M. r. = 25.3 kg/mm^2 at 750°	Sintered
NiAl	CsCl	1650	M. r. = 91.5 kg/mm^2; work of impact 17.3 kg/cm	M. r. = 112.5 kg/mm^2 at 1100°; σ_{100} at 1000° 8.45 kg/mm^2	With additions of Ti and (or) Zr
(Fe, Ni)Al	CsCl	∼1400	M. r. = 47.8 kg/mm^2	—	—
TiAl	Tetr. (CuAu)	1460 *	$\sigma_b = 28.1$ kg/mm^2; $\delta_{cont} = 10–15\%$; work of impact 4.6–6.9 kg/cm	—	—
TiAl	Tetr. (CuAu)	1460 *	M. r. = 83.0 kg/mm^2	M. r. = 70.3 kg/mm^2 at 1000° and 35.2 at 1100°; σ_{20} at 950° 8.45 kg/mm^2	—
TiAl+NiAl		—	M. r. = 15.5 kg/mm^2	—	—
TiAl+Ni		—	M. r. = 52 kg/mm^2	—	—
TiNi	CsCl	1310°	$\sigma_b = 80–95$ kg/mm^2; $\delta = 15\%$; $a_n = 3.8$ kg-m/cm^2	$\sigma_b = 87$ kg/mm^2 at 649°	—
Mo aluminides	Mixed	<1800	M. r. = 24.6 kg/mm^2	M. r. = 56.7 kg/mm^2 at 1100°	—
Cr_2Ti	Laves TC14 and TC15	1350**	M. r. = 43.5 kg/mm^2	92.5 kg/mm^2 at 1100°	—
Cr_2Ti	Same	1350 **	M. r. = 43.5 kg/mm^2	At 1000°: $\sigma_{50} = 13.0$ kg/mm^2; $\sigma_{100} = 12.45$; $\sigma_{500} = 10.25$ kg/mm^2	—
Ni_4Zr	—	1600	—	—	May form in the cold
Co—Cr—Mo R-phase	Complex	— —	— —	M. r. = 26.7 kg/mm^2 at 1000°	—
σ-phase	Compl. cub.	—	—	M. r. = 19.0 kg/mm^2 at 1000°	—
Co—Cr σ-phase	Compl. cub.	∼1450	—	M. r. = 20.4 kg/mm^2 at 1000°	—

*Peritectic temperature.   **Disordering temperature; m.p. ∼1400°. σ_{100} — long-term tensile strength after 100 hr, in kg/mm^2.   M.r. —modulus of rupture.

and TiAl, the superconductor compounds Nb_3Al, Zr_3Al, and $ThAl_3$, etc. Some aluminides, enriched in aluminum atoms ($NiAl_3$, $TiAl_3$, $ZrAl_3$, etc.), play an important role in strengthening alloys based on light metals.

Data on the structures and certain properties of aluminides, taken from various literature sources [226], are given in Table 36.

Some Properties of Semiconductor Compounds

Those metallides which we classify as semiconductor compounds have special physical properties. They differ from ordinary intermetallides in having high electrical resistivity (see Fig. 83b), negative coefficients of electrical resistivity, low thermal conductivity, and high hardness. The most characteristic property of semiconductor compounds is the forbidden-band width. Certain rules govern the variation of optical and dielectric constants, current-carrier mobility, amplitude of atomic vibrations, and catalytic and other properties of semiconductor compounds.

All these special physical and mechanical properties are discussed in detail in [105]. The cited book is valuable in this respect, that these properties are considered in it from the chemical viewpoint — the effect of the relative disposition of the reacting elements in the periodic system and the resulting electron redistribution in the structure of the semiconductor compounds.

Emphasizing the complexity of variation of these properties within subgroups of elements, the author of [105] establishes certain rules governing it. Thus, for instance, the hardness of the phosphides InP, GaP, AlP, and BP (aluminum phosphide excepted) increases in that order from InP to BP, i.e., with increasing difference in electronegativity of their components. Their resistivity and, especially, forbidden-band width vary in the same order.

Some data on the dependence of forbidden-band width ΔE for a number of semiconductor phosphides, arsenides, and antimonides of aluminum, gallium, and indium were given in Fig. 15. These data show that the forbidden-gap width decreases in that order on passing from aluminum phosphide to gallium and indium phosphides. In this case the interatomic bonds become progressively less directional (covalent) as ΔE decreases. This corresponds to gradual "metallization" of the chemical bonding of such semiconductor compounds owing to weakening of electron-pair bonds and gradual transition of the electrons to the free-electron gas in the structure of the compounds. Complete transition of such electrons to the free or semifree state attests to transition from covalent to metallic bonding.

Data taken from [105] on some physicochemical and physical properties of semiconductor compounds having the diamond structure, formed by A_{III} and B_V elements, are given in Table 37.

Using these data, one can follow the properties of such individual substances and the order of their variation with respect to the relative disposition of their components in the periodic system.

Recently the optical properties of semiconductor compounds have become quite important in connection with the possibility of using them in the latest quantum generators, called lasers. Gallium arsenide, GaAs, played a prominent role in this connection. The light generator based on it provides high radiation efficiency (90-100% versus 1-10% for earlier generators of ruby containing additions of chromium).

From the viewpoint of the theory and practical application of semiconductor compounds, the variation of a number of physical properties of semiconductor-metallide systems with composition is very interesting. Electrical, thermal, mechanical, and other properties are important in this respect. It will be noted that the different properties of such systems vary in accordance with the laws of physicochemical analysis established earlier by Kurnakov.

One of the characteristic properties of semiconductor compounds, as noted above, is their forbidden-band width. The variation of this property in individual compounds was considered above. There are some literature data showing the rules governing the variation of these properties with composition and structure for systems of semiconductor compounds.

TABLE 35. Some Properties of Titanium, Zirconium, Niobium, Tantalium, and Molybdenum Beryllides

Compound	Pressing No.	M. p., $^{\circ}$C	Sample density		Test temperature, $^{\circ}$C	Tensile strength, kg/mm^2	Modulus of elasticity, kg/mm^2	Remarks
			g/cm^3	% of abs. density				
TiBe$_{12}$	153	1535	2.29	99.1	1260	6.4	—	—
TiBe$_{12}$	153		2.29	99.1	1260	9.14	4920	—
TiBe$_{12}$	153		2.29	99.1	1370	2.46	—	—
TiBe$_{12}$	153		2.29	99.1	1370	2.46	—	—
TiBe$_{12}$	153		2.29	99.1	1520	1.62	—	—
TiBe$_{12}$	153		2.29	99.1	1520	2.1	—	—
TiBe$_{12}$	48		2.29	100	1260	3.87	—	—
TiBe$_{12}$	48		2.29	100	1260	3.1	—	—
ZrBe$_{13}$	96	1900	2.78	98.6	1260	13.0	—	3.2% BeO
ZrBe$_{13}$	96		2.79	99.0	1370	10.6	—	0.5% Al
ZrBe$_{13}$	96		2.78	98.6	1520	8.84	—	—
ZrBe$_{13}$	109		2.75	98.5	1260	25.9	—	2.2% BeO
ZrBe$_{13}$	109		2.75	98.5	1370	26.3	—	0.15% Fe
ZrBe$_{13}$	109		2.74	98.2	1520	17.3	—	—
ZrBe$_{13}$	185		2.61	94.4	1260	20.6	28100	—
ZrBe$_{13}$	185		2.62	94.5	1370	19.75	14050	—
ZrBe$_{13}$	185		2.61	94.2	1370	22.2	11950	0.84% BeO
ZrBe$_{13}$	185		2.59	93.5	1520	11.5	7030	0.08% Fe
ZrBe$_{13}$	41		2.72	96.5	1260	12.1	—	2.0% BeO
ZrBe$_{13}$	41		2.72	96.5	1370	5.28	—	0.02% Al
ZrBe$_{17}$	143	>1920	3.04	99.4	1260	31.1	16900	—
ZrBe$_{17}$	143		3.04	99.4	1260	24.5	18300	—
ZrBe$_{17}$	143		3.05	99.6	1370	20.4	10550	—
ZrBe$_{17}$	143		3.04	99.4	1370	17.8	9850	—
ZrBe$_{17}$	143		3.05	99.6	1520	17.2	7030	—
NbBe$_{12}$	34	1700	2.88	99.0	1260	7.4	—	—
NbBe$_{12}$	34		2.86	98.4	1260	6.6	—	—
NbBe$_{12}$	59		2.83	97.3	1370	27.6	—	—
NbBe$_{12}$	59		2.91	100.0	1485	>28.6	—	Not decompd.
NbBe$_{12}$	100		2.86	97.2	1260	27.6	—	—
NbBe$_{12}$	100		2.86	97.2	1370	27.6	—	—
NbBe$_{12}$	100		2.85	97.0	1520	13.15	—	—
NbBe$_{12}$	101		2.72	91.9	1260	23.9	—	—
NbBe$_{12}$	101		2.71	91.5	1370	23.7	—	—
NbBe$_{12}$	101		2.71	91.5	1520	>16.2	—	—
NbBe$_{12}$	108		2.77	98.2	1260	27.3	—	—
NbBe$_{12}$	108		2.78	98.6	1370	23.0	—	—
NbBe$_{12}$	108		2.74	97.1	1520	13.3	—	—
NbBe$_{17}$	179	>1700	3.15	98.5	1260	27.4	15470	—
NbBe$_{17}$	179		3.15	98.5	1370	>26.4	9850	—
NbBe$_{17}$	179		3.14	98.1	1520	>11.5	—	—
NbBe$_{11}$	141	—	3.00	100	1260	42.5	16900	—
NbBe$_{11}$	141		2.92	97.5	1260	36.0	16200	—
NbBe$_{11}$	141		2.99	99.6	1370	42.5	—	—
NbBe$_{11}$	141		2.86	95.4	1370	32.7	12660	—
NbBe$_{11}$	141		2.97	99.0	1520	>22.9	—	—

TABLE 35
(continued)

Compound	Pressing No.	M. p., °C	Sample density		Test temperature, °C	Tensile strength, kg/mm^2	Modulus of elasticity, kg/mm^2	Remarks
			g/cm^3	% of abs. density				
TaBe$_{12}$	149	1850	4.07	96.0	1260	34.4	16900	—
TaBe$_{12}$	149		4.07	96.0	1260	40.5	16900	—
TaBe$_{12}$	149		4.07	96.0	1370	30.2	9850	—
TaBe$_{12}$	149		4.07	96.0	1370	20.4	9150	—
TaBe$_{12}$	149		4.07	96.0	1520	18.65	6230	—
TaBe$_{12}$	149		4.07	96.0	1520	18.65	7030	—
TaBe$_{12}$	72		4.12	97.2	1485	4.15	—	—
TaBe$_{12}$	72		4.11	97.0	1485	4.9	—	—
Ta$_2$Be$_{17}$	178	—	4.45	97.3	1260	39.6	18300	—
Ta$_2$Be$_{17}$	178		4.44	97.1	1260	40.4	—	—
Ta$_2$Be$_{17}$	178		4.43	97.0	1370	>26.4	9150	Not decompd.
Ta$_2$Be$_{17}$	178		4.44	97.1	1520	>13.15	—	Not decompd.
TaBe$_{11}$	176	—	4.20	98.8	1260	33.1	18300	—
TaBe$_{11}$	176		4.21	99.1	1370	26.4	7740	—
TaBe$_{11}$	176		4.19	98.6	1370	26.7	9150	—
TaBe$_{11}$	176		4.20	98.8	1520	>13.15	∾4200	—
MoBe$_{12}$	148	1700	2.96	95.8	1260	29.4	10550	—
MoBe$_{12}$	148		3.02	97.7	1370	21.2	8450	—
MoBe$_{12}$	148		3.02	97.7	1370	20.8	7740	—
MoBe$_{12}$	148		3.02	97.7	1520	8.85	700	—
MoBe$_{12}$	148		3.02	97.7	1520	7.74	—	—

Thus, for instance, it was found that in the systems of continuous solid solutions AlSb−GaSb and possibly InP−InAs, the forbidden-gap width varies almost linearly with composition. The variation of this property for these two systems is shown in a forbidden-gap width−composition diagram plotted from data of [240] (see Fig. 84). In systems with limited solid solutions the forbidden-gap width changes abruptly. Some data on these properties are given in [105].

Thus, for instance, it is reported that for the system InTe−Ga$_2$Te$_3$, there is not only a solubility gap, but also a break in the curve of forbidden-gap width (ΔE) versus composition. The break extends from 55 to 73 mole % Ge$_2$Te$_3$ concentration, and this region corresponds to the two-phase part of the phase diagram for the system InTe−Ga$_2$Te$_3$.

Analogous breaks in the ΔE−composition curves for semiconductor systems may appear when ternary compounds are formed in these systems. Such compounds may result from reactions between binary metallides, as is shown by the formation of the ternary compounds ZnGa$_2$Se$_4$ in the system ZnSe−Ga$_2$Se$_3$, CdIn$_2$Se$_4$ in the system CdSe−In$_2$Se$_3$, CdIn$_2$Te$_4$ in the system CdTe−In$_2$Te$_3$, CdGa$_2$S$_4$ in the system CdS−Ga$_2$S$_3$, etc.

Similar rules govern variations in hardness, thermal conductivity, electrical, and other properties of systems of semiconductor compounds.

With regard to the dependence of all these properties on composition and structure in a system of semiconductor metallides, the system Bi$_2$Se$_3$−Bi$_2$Se$_3$ with limited solid solutions and ternary metallides [181] is of

TABLE 36. Compositions and Properties of Some Aluminides

Compound	Structure type	Heat of formation, kcal/mole	M.p., °C	Electrical resistivity, $\mu\Omega\text{-cm}$	Hardness, kg/mm^2
$ThAl_3$	Hex.	—	—	—	—
$ThAl_2$	Hex.	—	—	—	—
Th_3Al_2	Tetrag.	—	—	—	—
Th_2Al	Tetrag. $(CuAl_2)$	—	—	—	—
$ZrAl_3$	Tetrag.	—	—	—	—
Zr_3Al	Cub. (Cu_3Na)	—	—	—	$H_5=445$
Zr_3Al_2	—	—	—	—	$H_5=475$
Zr_5Al_3	—	—	—	—	$H_5=580$
$ZrAl_2$	Rhomb.	—	—	—	—
$TiAl_3$	f. c. tetrag.	8.75	1400	—	$H_5=443,73$ and $334,0$
$TiAl$	—	9.6	1520	—	$H_5=156,25$ *
Ta_5Al_3	Tetrag.	—	—	—	$H_3=440 \div 450$
Nb_3Al	Cub. $(\beta\text{-}W)$	—	—	—	—
Nb_2Al	σ-phase	—	—	300	—
$NbAl_3$	Tetrag. $(TiAl_3)$	—	—	—	$H_5=375$
VAl_3	f.c. tetrag. $(TiAl_3)$	—	—	—	$H_R=82$
VAl_5	—	—	—	—	$H_R=125$
VAl_6	Hex.	—	—	—	$H_R=50 \div 116$
VAl_7	—	—	—	—	$H_R=45 \div 110$
VAl_{11}	f.c.c.	—	—	—	$H_R=40$
V_5Al_8	f.c.c. $(\gamma\text{-brass})$	—	1590	—	—
UAl_3	Cub. $(CuMg)$	—	1350	—	—
UAl_2	Cub. (Cu_2Mg)	—	730	—	—
UAl_5	—	—	—	—	—
WAl_5	Hex.	—	—	—	—
WAl_{12}	Cub.	—	—	—	$H_5=536$
$MoAl_5$	Hex.	—	—	—	$H_5=366$
$MoAl_{12}$	Cub.	—	—	—	—
$CrAl_7$	Orthorh.	—	—	—	$H_5=510$**
Cr_2Al_{11}	Orthorh.	—	—	—	$H_5=710$
$PtAl_2$	Cub.	—	—	—	—
$FeAl_3$	Orthorh.	6.7	—	100	$H_5=960$
Fe_2Al_5	—	6.4	—	—	—
$FeAl_2$	—	6.5	—	—	—
$FeAl$	b.c.c.	6.1	—	—	—
Co_2Al_9	Monocl.	38.5	—	—	—
$CoAl_4$	—	—	—	—	$H_5=735$
Co_2Al_5	Hex.	70.0	—	—	—
$CoAl$	Cub. $(CsCl)$	26.4	1628	—	$H_5=530$
$NiAl$	Cub. $(CsCl)$	34.0	1640	—	$H_5=500$
$NiAl_2$	—	38.7	—	16.6	—
$NiAl_3$	Orthorh.	38.0	—	—	—
Ni_2Al_3	Rhombohedr.	—	—	—	$H_5=770$
Ni_3Al	Cub. (Cu_3Au)	37.6	1362	16.35	—

*Modulus of elasticity $\times 10^3 = 14.41$ kg/mm^2; tensile strength 83 kg/mm^2.

** Tensile strength 87 kg/mm^2 at 500°.

TABLE 37. Physical Properties of Analogs of Semiconductor Compounds

Compound	Lattice parameter, A	M. p., °C	Microhardness, kg/mm²	ΔE, eV	Electron mobility, cm²/V-sec	Coefficient of linear expansion, $\alpha \cdot 10^6$ deg⁻¹	Refractive index	Dielectric constant	Thermal conductivity, cal/cm-sec-deg
BN_w	a=2.504, c=6.661	—	—	4.6——3.6	—	10.5	—	4.15	at 300° 0.036
BN_s	3.615	—	—	—	—	7.5	—	—	—
BP	4.537	—	3200	5.9	—	—	3.0	—	—
BAs	4.777	—	—	—	—	—	—	—	—
AlN_w	a=3.104, c=3.965	2200	—	3.8	—	—	—	—	—
AlP	5.45	—	—	3.0	—	—	—	—	—
AlAs	5.6622	>1600	500±20	2.16	—	3.5	—	—	—
AlSb	6.1355	1060	400±20	1.6	200	—	3.0	—	—
GaN_w	a=3.180, c=5.160	1500	—	3.25	—	—	—	—	—
GaP	4.4505	1350	940±35	2.4	100	3.5	2.9	—	—
GaAs	5.6534	1237	700±50	1.53	6000	5.8	3.2	11.1	0.125
GaSb	6.0954	703	420	0.80	4000	6.9	3.7	—	0.105
InN_w	a=3.53, c=5.69	—	—	2.4	—	—	—	—	—
InP	5.8687	1070	435±20	1.34	3400	—	3.0	—	—
InAs	6.0584	942	330±10	0.45	23000	5.3	3.2	—	—
InSb	6.47877	536	220±10	0.25	65000	5.5	4.1	—	0.037

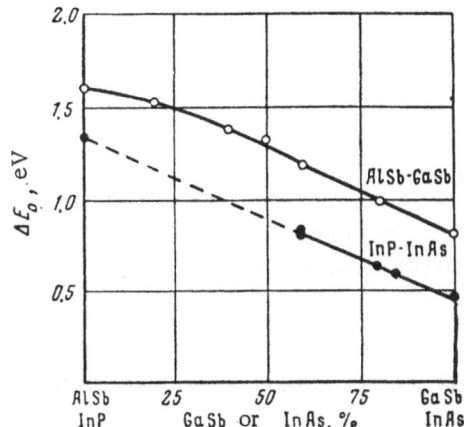

Fig. 84. Curves of forbidden-band width (obtained by extrapolation to 0°K) versus composition of solid solutions in systems AlSb—GaSb and InP—InAs.

interest. The authors of this paper described the phase diagram shown in Fig. 85; it characterizes a solubility gap in the system (solubility limits are shown in the diagram by dashed lines owing to the insufficiency of experimental data). The solubility gap in this binary system is due to the difference in type of structure between bismuth selenide, which has a rhombohedral, tridymite-type structure (C33), and bismuth sulfide, which has an orthorhombic structure (type D-5₈).

Investigation of electrical and thermal properties, as well as hardness, in this system revealed very interesting relations between these properties and composition and structure; this enables one to discover new phenomena connected with chemical affinity in metallide solid solutions.

The fusibility diagram of the system $Bi_2Se_3 - Bi_2S_3$ is shown in Fig. 85 at the top, curves of electrical conductivity lg σ and thermo-emf α versus composition in the middle, and microhardness H_μ and thermal conductivity λ versus composition at the bottom.

Analysis of these curves shows that in the region of limited solid solutions based on bismuth selenide, the properties lg σ and α vary continuously; breaks in the curves appear at the solubility limits. Within the region of existence of limited solid solutions based on Bi_2S_3, the curves of log σ and α each have one singular point which corresponds to the ternary compound Bi_2SeS_2. It also corresponds to the electrical conductivity maximum for this compound.

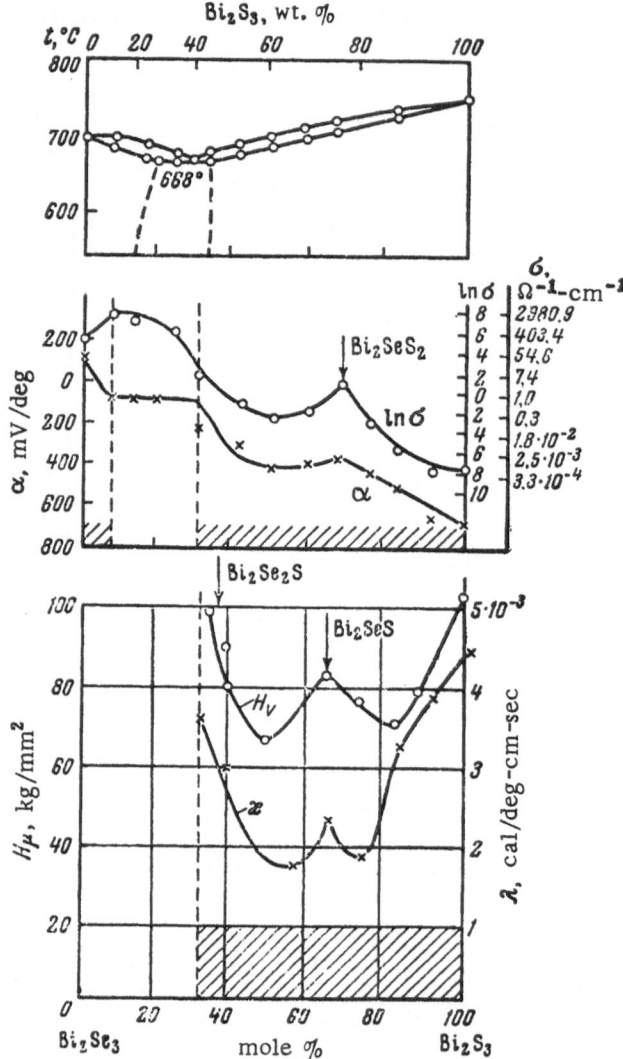

Fig. 85. Diagrams of fusibility, thermo-emf α, electrical conductivity σ, microhardness H_μ, and thermal conductivity λ for the system $Bi_2Se_3-Bi_2S_3$.

The maximum value of this property is apparently connected with the increased current-carrier mobility due to the ordered structure of this compound.

Although the formation of the ternary compound Bi_2SeS_2 in this system is not discussed in [181], it presumably results from transformation of the solid solution based on Bi_2S_3. In this connection, Bi_2SeS_2 is a classic example of a ternary Kurnakov compound formed by reactions of two semiconductor-type metallides in limited solid solutions.

Microhardness and thermal conductivity data given in Fig. 85 confirm the presence of Bi_2SeS_2. More-over, the authors, based on maximum microhardness and thermal conductivity values corresponding approxi-mately to the solubility limit of Bi_2Se_3 in Bi_2S_3, assume the possibility of formation of a second compound, Bi_2Se_2S. Although this assumption is less well grounded than that of formation of the first compound, the

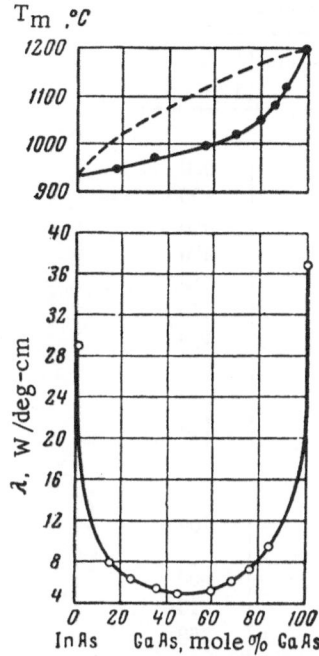

Fig. 86. Diagrams of fusibility and thermal conductivity λ versus composition for system InAs—GaAs.

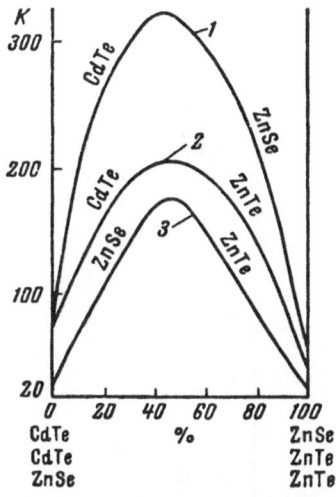

Fig. 87. Diagrams of thermal resistivity K versus composition for systems CdTe—ZnSe, CdTe—ZnTe, and ZnSe—ZnTe: 1) ZnSe—CdTe; 2) ZnTe—CdTe; 3) ZnSe—ZnTe.

second is shown in Fig. 85 as actually existing. The authors prove the presence of these two compounds not only by data on singular points in the property—composition diagrams, but also by studying the crystal structures of the compounds, particularly the ordered arrangement of sulfur and selenium atoms, characteristic of them.

At present, important rules have been established governing the variation of current-carrier mobility, thermal conductivity, and other physical properties in systems of semiconductor compounds [105, 240-242]. The theory of the thermal conductivity of semiconductors, including semiconductor systems, and its dependence on composition and structure is set forth in a monograph [241], whereas questions of thermal conductivity in solid solutions of semiconductors are discussed in [242].

Thermal conductivity is especially important for understanding many physical phenomena in semiconductor systems, connected with current-carrier mobility, the mechanism of heat transfer in solids, atomic vibrations in crystal structures, etc. Practically they are important for thermo-element systems; this topic is discussed in [241, 242].

Today thermal conductivity has been studied in many semiconductor systems, in most of which continuous solid solutions are formed: InAs—InP, InAs—GaAs, GaAs—GaP, PbTe—PbSe, HgTe—HgSe, Bi_2Te_3—Bi_2Se_3, Bi_2Te_3—Sb_2Te_3, and PbTe—SnTe [241].

In such systems with continuous solid solutions, the thermal conductivity—composition curve is continuous and passes through a minimum corresponding to about 50 mole % of one of the compounds. A striking example of this behavior of thermal conductivity appears in the thermal conductivity—composition diagram of the system InAs—GaAs, shown in Fig. 86. As is evident from the top of the figure, this is a system of continuous solid solutions, the fraction of electronic thermal conductivity λ being negligibly small in this case; only the structure thermal conductivity and its variation with composition are shown in the property—composition diagram.

As experimental data show, at a content of about 50 mole % GaAs in solid solutions of this system, the thermal conductivity is about 1/30 of that of pure InAs, or 1/36 of that of pure GaAs. This rule of variation in thermal conductivity applies to all systems consisting of solid solutions of semiconductor compounds [242]. In this paper the thermal conductivity of solid solutions of semiconductor compounds was investigated experimentally and equations proposed describing the dependence of this property on component concentration in such systems; the authors introduced for calculation the inverse thermal conductivity $1/\chi$, calling it the thermal resistivity K. To express the thermal resistivity of solid solutions of semiconductors versus composition, they proposed the formula

$$K = K_1 C + K_2 (1 - C) + (S_1 + S_2) C (1 - C),$$

where K is the thermal resistivity of a solution having the concentration C; K_1 and K_2 are the thermal resitivitites of the components; S_1 and S_2 are quantities characterizing the scattering introduced by each component.

TABLE 38. Composition and Some Properties of Ferromagnetic Compounds

System	Composition	Structure type	Temperature of formation, °C	Curie point °C
Ferromagnetic Kumakov compounds [43,227,228]				
Fe—Al	Fe$_3$Al	Ord. cub.	580	500
Fe—Cr	FeCr	Dist. tetrag.	920	100
Fe—Co	FeCo	Ord. cub. (CsCl)	730	980
Fe—N	Fe$_4$N	Ord. face-cent.	680	490
Fe—Ni	FeNi$_3$	Ord. cub.	503	615
Fe—Pd	FePd	Ord. tetrag. (AuCu)	700	470
Fe—Pt	FePt	Ord. tetrag. (AuCu)	1320	480
Fe—Si	Fe$_3$Si	Ord. cub. (BiFe$_3$)	~1250	~600
Co—Pt	CoPt	Ord. tetrag. (AuCu)	825	580
Mn—N	Mn$_4$N	Ord. face-cent.	~700	465
Ferromagnetic compounds crystallized from the melt [227,228]				
Fe—B	Fe$_2$B	Tetrag. (CuAl$_2$)	Peritect. 1389	769
Fe—Be	FeBe$_2$	Hex. (MgZn$_2$)	1420	650
Fe—C	Fe$_3$C	Orthorh.	Peritect	215
Fe—Ce	CeFe$_2$		Peritect 1060	768
Fe—P	Fe$_3$P	Tetrag.	1166	420
	Fe$_2$P	Hex.	1365	420
Fe—S	FeS	(NiAs)	1190	320
Co—B	Co$_2$B	Tetrag. (CuAl$_2$)	1265	510
Mn—As	MnAs	Hex. (NiAs)	935	45
Mn—B	MnB	Orthorh.	—	562
Mn—Bi	MnBi	Hex. (NiAs)	Peritect 1043	445
Mn—P	Mn$_2$P	Hex.	1327	40
	MnP	Orthorh.	1147	317
Mn—Sb	Mn$_2$Sb	Tetrag. (Cu$_2$Sb)	948	277
	MnSb	Hex. (NiAs)	Peritect 872	314
Mn—Sn	Mn$_3$Sn	Hex. (Ni$_3$Sn)	Peritect 989	117
	Mn$_2$Sn	Hex. (NiAs)	Peritect 897	263
Ferromagnetic Heusler alloys [228,231]				
Mn—Cu—Al	MnCu$_2$Al	Ord. β-phase	—	630°K
Mn—Cu—In	MnCu$_2$In	b. c. c. (Fe$_3$Al)	—	500°K
Mn—Cu—Ga	MnCu$_2$Ga	b. c. c.	—	—
Mn—Cu—Ge	MnCu$_2$Ge	tetrag. $c/a=0,96$	—	300°K
Mn—Cu—Sn	MnCu$_2$Sn	Ord. β- phase cub. (Fe$_3$Al)	—	—
Mn—Cu—Al	MnCu$_2$Al$_2$	Same	—	—
Mn—Cu—Bi	MnCu$_2$Bi	—	—	200°K
Ferromagnetic compounds of rare-earth metals [232,233]				
Pr—Ru	PrRu$_2$	Cub. (MgCu$_2$)	—	40°K
Pr—Rh	PrRh$_2$	Same	—	8.6°K
Pr—Os	PrOs$_2$	»	—	35°K
Pr—Ir	PrIr$_2$	»	—	18.5°K
Fr—Pt	PrPt$_2$	»	—	7.9°K

TABLE 38 (continued)

System	Composition	Structure type	Temperature of formation, °C	Curie point, °C
Pr—Si	PrSi$_2$	»	—	35°K
Nd—Ru	NdRu$_2$	»	—	10,5
Nd—Rh	NdRh$_2$	»	—	8.1°K
Nd—Ir	NdIr$_2$	»	—	11.8°K
Nd—Pt	NdPt	»	—	6.7°K
Gd—Ru	GdRu$_2$	»	—	>77°K
Gd—Rh	GdRh$_2$	Hex. (MgZn$_2$)	—	>77°K
Gd—Os	GdOs	Same	—	>77°K
Gd—Ir	GdIr$_2$	Cub. (MgCu$_2$)	—	>77°K
Gd—Pt	GdPt$_2$	Same	—	77°K
Sm—Os	SmOs$_2$	Hex. (MgZn$_2$)	—	34°K
Er—Ru	ErRu$_2$	Hex. (MgZn$_2$)	—	13°K
Pt—Ni	PtNi$_2$	Cub. (MgCu$_2$)	—	8°K
Nd—Ni	NdNi$_2$	—	—	20°K
Sm—Ni	SmNi$_2$	—	—	77°K
Gd—Ni	GdNi$_2$	Cub. (MgCu$_2$)	—	90°K
Tb—Ni	TbNi$_2$	—	—	46°K
Dy—Ni	DyNi$_2$	Cub. (MgCu$_2$)	—	32°K
Ho—Ni	HoNi$_2$	—	—	23°K
Er—Ni	ErNi$_2$	—	—	14°K
Tm—Ni	TmNi$_2$	—	—	14°K

Mechanical mixtures of semiconductor systems are characterized by the linear relation expressed by the simple formula

$$K = K_1 C + K_2 (1 - C).$$

These equations were confirmed in a study of thermal conductivity in several systems of semiconductor compounds.

Curves of thermal resistivity K versus composition are shown in Fig. 87 for systems with continuous solid solutions CdTe—ZnSe, ZnTe—ZnSe, and CdTe—ZnTe. In accordance with theory, their curves of thermal resistivity K versus composition all pass through a flat maximum corresponding to about 50 mole % concentration of one of the components.

The equations for calculating the thermal resistivity of semiconductor metallide systems are of value in studying the general rules governing this property with respect to the character of interaction and type of phase diagram of each system. They are important also in the design of thermoelectric batteries.

Magnetic Properties of Metallides

Many metallides, especially those formed by ferromagnetic components (Fe, Co, Ni), have magnetic properties. These properties are sensitive to phase transitions connected with ordering of structure and separation of phases. The interesting groups of ferromagnetic metallides include Kurnakov compounds, which are formed, as noted above, by transformation of solid solutions. The compounds Fe$_3$Al, FeNi$_3$, MnNi$_3$, FeCo, FePt, etc., are in this group. A second group of such compounds comprises the so-called Heusler magnetic alloys: Cu$_2$MnAl, Cu$_2$MnSn, Cu$_2$MnGa, etc.

In the first group of ferromagnetic compounds, at least one of the components is ferromagnetic; in the second group, none of the components are ferromagnetic; instead, they are paramagnetic and diamagnetic metals as a rule. The ferromagnetic properties of such metallides result from the special character of chemical reactions among the components, phase transitions in the solid state, and redistribution of electrons in the atoms of the crystal structure.

The degree of ordering of structure in these alloys is important in imparting pronounced magnetic properties to them. Hence in order to learn the compositions and properties of such compounds one must turn to the equilibrium diagrams of systems containing compounds with ordered structures. These questions are discussed in detail in special monographs [227, 228] and separate collections and articles [229, 230]. Hence the many compositions and properties of such compounds need not be given here. In order to become generally familiar with the individual characteristics of a ferromagnetic compound, one requires data on their compositions, structure type, temperatures of formation, and ferromagnetic properties as determined from their Curie points.

Data for a number of compounds are given in Table 38 [43, 227-230].

An investigation of the properties of compounds of the type of Heusler alloys is described in a recent paper [231], which is our main source of data on crystal structures and Curie points. In this paper it is shown that the ferromagnetic properties of these metallides increase with the degree of ordering in them and the extent to which the characteristic arrangement of manganese atoms Mn—Mn at definite, shortened distances occurs in their structure. The exact atomic ratio and the conditions of heat treatment are very important in determining the ferromagnetic properties of these compounds. The maximum values of magnetic properties and Curie points of alloys of the system Mn—Cu—Al correspond exactly to the composition Cu_2MnAl. The magnetic properties are enhanced after quenching and subsequent low-temperature aging (at 110-200°).

In [231] it is shown also that the ferromagnetic compound $MnCu_2Sb$, reported earlier [227, 228], is not such. At room temperature it is a paramagnetic material, whereas on cooling below 38° K it goes over to an antiferromagnetic. According to data of the same investigation, contrary to earlier reports, $MnCu_2Bi$ and $MnCu_2Ga$ are polyphase alloys. The ferromagnetism of these alloys apparently is connected with decomposition of the alloys in the solid state and the effect of binary phases on these properties.

The tabulated data on the ferromagnetic properties of metallides containing rare-earth metals were taken from [232, 233]. The magnetic properties of nickel compounds containing these metals were studied in [233]. As is evident from Table 38, all these compounds are paramagnetic at room temperature, whereas ferromagnetic properties are acquired only at subzero temperatures. Gadolinium compounds have the highest Curie points.

One should note the development of investigations in the field of the magnetic properties and structure of rare-earth metal compounds with transition metals, and the relation between the magnetic properties and structure of Cu_5Ca-type metallides [234-236]. In these papers it was shown by x-ray structure analysis and other methods that many compounds of this type, composed of these elements, have the same Cu_5Ca-type hexagonal structure with characteristic arrangement of the rare-earth and transition metal atoms.

The isomorphous hexagonal structures of the following compounds: Co_5Y, Co_5Ce, Co_5P_2, Co_5Nd, Co_5Gd, Co_5Dy, Co_5Er, Ni_5Y, Ni_5La, Ni_5Ce, Ni_5P_2, Ni_5Nd, Ni_5Gd, Ni_5Dy, Ni_5Er, Fe_4Ho, Fe_4Er, Fe_4In, Cu_5Y, Cu_5Nd, and Cu_5Gd, were established in [236] and important data obtained on their magnetic moments. The rules governing their variation with respect to the electronic structure of the reacting components were established, as well as the difference in properties between nickel compounds and the analogous cobalt and iron ones.

Oxidation Resistance of Refractory Metallides in Air

In connection with the properties of metallides, their chemical stability in various media is of considerable interest. This subject is discussed in detail in the literature [237, 238]. The most complete information is given in a handbook [219].

TABLE 39 High-Temperature Oxidation Resistance of Refractory Metallides in Air

Compound	Temperature, °C	Oxidation time, h	Oxidation rate, mg/cm²−h	Compound	Temperature, °C	Oxidation time, h	Oxidation rate, mg/cm²−h
TiB_2	1000	1	+12.0	Cr_3C_2	1100	4	0
TiB_2	1000	170	+0.18	WC	1000	1	+13.7
TiB_2	1200	2	+5.0	TiN	1000	1	+25.0
TiB_2	1200	50	+1.24	$TiSi_2$	1260	100	+0.023
TiB_2	1200	100	+0.74	$ZrSi_2$	1100	4	+9.5
ZrB_2	1000	150	+0.2	V_3Si	1250	1	−63.0
ZrB_2	1150	8	+0.06	VSi_2	1200	4	+1.2
ZrB_2	1150	200	+0.02	$NbSi_2$	1200	4	−13.5
NbB_2	1000	1	+32.5	$TaSi_2$	1500	1	+2.0
TaB_2	900	2	+1.67	Cr_3Si	1300	4	+2.9
CrB_2	1000	150	+0.014	$CrSi_2$	1300	4	+2.2
TiC	1000	1	+0.8	$CrSi_2$	1300	12,2	−0.11
TiC	1100	2	+4.81	Mo_3Si	1500	4	−203.0
TiC	1150	48	Decomposes	Mo_5Si_3	1500	4	−16.9
ZrC	1100—1400	Beginning of active oxidation	—	$MoSi_2$	1155	200	+0.04
VC	900	—	+73.5	$MoSi_2$	1500	4	+0.32
NbC	900	—	+20.5	$MoSi_2$	1320	50	+5.0
NbC	1100—1400	Beginning of active oxidation	—	$MoSi_2$	1320	100	+4.0
TaC	900	2	+19.7	$MoSi_2$	1565	100	−3.67
TaC	1100—1400	Beginning of active oxidation	—	$MoSi_2$	1565	135	−3.10
$Cr_{23}C_6$	1000	1	0	WSi_2	1500	4	−5.9
$Cr_{23}C_6$	1100	1	0	$MnSi_2$	1200	4,0	+1.7
Cr_7C_3	1000	1	+35.0	$MnSi_2$	1200	6,0	+1.25
Cr_3C_2	800	1	0	$ReSi_2$	1400	0,5	+6.4
Cr_3C_2	1000	1	0	$FeSi_2$	1200	1,0	+0.4

One important chemical property of metallides is their resistance to oxidation in air when hot (high-temperature oxidation resistance). The fact that many compounds whose components (boron, carbon, silicon, molybdenum, titanium, etc.) are very easily oxidized on heating have very high-temperature oxidation resistance, is characteristic in this respect. For instance, molybdenum and silicon are very easily oxidized on heating, whereas the compounds formed between them — silicides (especially $MoSi_2$) — are very oxidation-resistant up to 1500-1600°.

Without analyzing the high-temperature oxidation resistance of refractory compounds in detail, we limit ourselves to giving data from a handbook [219], characterizing this property, in Table 39.

It is evident from Table 39 that silicides, borides, and carbides are most resistant to oxidation. Of these compounds, molybdenum, tungsten, and chromium disilicides, zirconium and titanium borides, and chromium carbides are exceptionally resistant; of all the compositions of chromium carbides, $Cr_{23}C_6$ and Cr_3C_6 are the most resistant and Cr_7C_3 the least. The diverse behavior of such compounds on heating in air is connected with the strength of interatomic bonding in the metallides and the nature of the oxide films. In some compounds these films protect the surface from oxidation, whereas in others they do not.

The data given in Table 39 on the oxidation resistance of refractory metallides play an important role in the practical use of these compounds in the form of various high-temperature oxidation-resistant materials

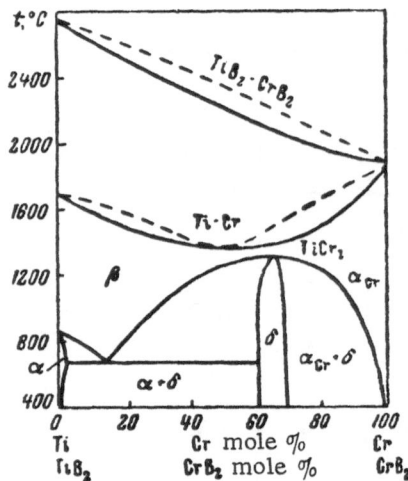

Fig. 88. Fusibility diagrams of systems Ti—Cr and TiB₂—CrB₂.

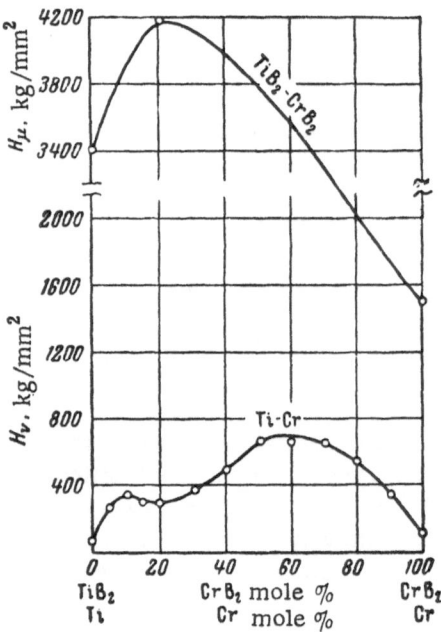

Fig. 89. Hardness-composition diagram of system Ti—Cr and microhardness— composition diagram of system TiB₂—CrB₂.

and oxidation-resistant coatings on refractory metals which are easily oxidized on heating (niobium, molybdenum, etc.).

Strength Properties of Metallide Systems

The intrinsic properties of metallides in equilibrium systems composed of them vary with composition and structure. These variations are usually governed by the same rules established by Kurnakov for ordinary metallic systems. However, the levels of physicochemical and mechanical properties are different; in many cases these properties have much higher values than in simple metallic systems. The rules governing the variation of properties with composition may be different in metallide systems where the component compounds have different types of chemical bonding. Considerable changes of properties in metallide systems should be expected also in processes of phase transition, ordering, etc.

In the consideration of individual metallide systems above, the variation of some of their properties with composition and structure was discussed. For instance, in connection with the properties of continuous solid solutions in the metallide system CrFe—VFe, formed from α-solid solutions of the system Fe—Cr—V, it was shown that hardness values for such binary solid solutions (σ-phases) are 3—4 times those for quenched ternary α-solid solutions having the same chemical compositions (see Fig. 19). Some other metallide systems differ considerably in properties from ordinary metallic systems. Here we shall discuss only two striking examples of the considerable enhancement of strength and heat resistance, caused by the formation of compounds and metallide solid solutions.

Fusibility diagrams of the binary systems Ti—Cr and TiB₂—CrB₂, composed of titanium and chromium and the borides of these metals, are given in Fig. 88. The first of these systems forms continuous solid solutions of chromium and β-Ti: One compound, TiCr₂, separating from β-solid solutions, undergoes eutectoid transformation on cooling in this system; in the other system, TiB₂—CrB₂, continuous solid solutions are formed between chromium and titanium borides without transformation in the solid state. The properties vary regularly in accordance with the character of the equilibrium diagrams of these systems.

Microhardness—composition curves for the systems Ti—Cr and TiB₂—CrB₂, plotted from literature data, are shown in Fig. 89. It is evident from comparison of these two curves that in the corresponding systems the pure components Ti and TiB₂, on the one hand, and Cr and CrB₂, on the other, have very different values of hardness. Titanium boride is about 30 times as hard as pure titanium, whereas chromium boride is 15-16 times as hard as pure chromium. This is due to the difference between the strength of bonding in the pure metals and that in their compounds.

In the system Ti—Cr hardness varies along a rather complex curve with a shallow, flat minimum in the region of eutectoid composition and a flat maximum in the region of the δ-phase based on TiCr₂. In the system of continuous solid solutions TiB₂—CrB₂ the microhardness varies along a smooth curve passing through

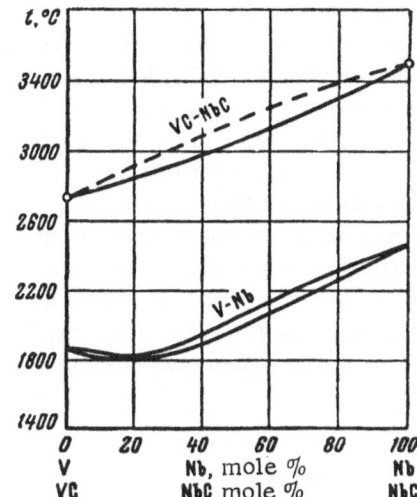

Fig. 90. Fusibility diagrams of systems V−Nb and VC−NbC.

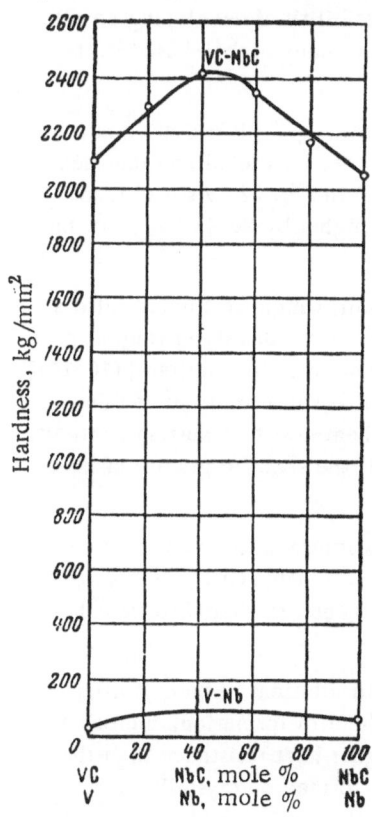

Fig. 91. Hardness−composition diagrams of systems V−Nb and VC−NbC.

a flat maximum of 20-30% CrB_2. In this case hardness values reach 4000-4200 kg/mm², which is about 40-42 times the hardness of pure titanium or chromium.

The individuality of properties of these compounds and their effect on the hardness of metallide solid solutions lie in the fact that the titanium boride is harder than chromium boride; the hardness of the solid solution based on titanium boride increases more markedly with concentration than that of the solution based on chromium boride. This is explained by the fact that the chemical bonding is stronger in TiB_2 than in CrB_2, since titanium boride is formed from components (Ti and B) which differ more in electronegativity than the components (Cr and B) of chromium boride. Compounds of this kind have larger heats of formation, higher melting points (see Fig. 88), and stronger chemical bonding. When the titanium atoms are replaced by chromium in the structure of solid solutions based on TiB_2, apparently the structure is distorted more and its hardness accordingly increased to a larger degree.

It is even more interesting to compare data on the variation in properties of solid solutions of the metal system V−Nb and those formed between vanadium and niobium monocarbides. In this case we have two analogous metals (V and Nb) and two carbides of these metals (VC and NbC). In accordance with the stronger chemical bonding in carbides compared with their component metals, the carbides have higher melting points. As a consequence, solid solutions of the system of these carbides also have higher incipient melting points than alloys of the vanadium−niobium system. This is clearly evident in the fusibility diagrams of the two systems (Fig. 90).

Solid solutions of the two analogous metals have relatively low hardness which varies along a smooth curve (Fig. 91). Maximum hardness values at 40-60% concentrations of the second component scarcely reach 100-120 units. The pure carbides (VC−NbC) are 30-40 times as hard as the pure metals. The hardness−composition curve of the monocarbide solid solutions also passes through a flat maximum. Maximum values are 2400-2450 kg/mm². This is 40-50 times the hardness of the pure constituent metals of these carbides.

In accordance with the fact that when vanadium and niobium carbides form mutual solid solutions, the metals exchanged (V and Nb) are analogs, the variation of hardness in this system is less marked than in the titanium boride−chromium boride system. Although the maximum hardness of monocarbide solid solutions exceeds that of vanadium monocarbide by 300-350 units, the difference between the the maximum hardness of the diboride solid solutions and that of titanium diboride is about 800 units.

Thus special conditions exist for strengthening metallide solid solutions, which are determined by the position of the interacting elements in the periodic system and their metallochemical properties. This explains the fact that exchange of Ti and Cr in diborides causes a larger increase in strength than exchange of V and Nb in monocarbide solid solutions.

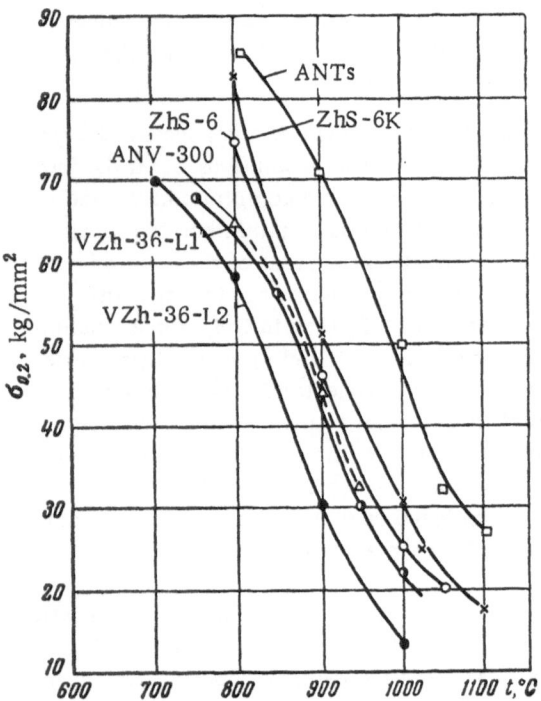

Fig. 92. Yield point of some heat-resistant alloys and an alloy based on metallide Ni₃Al (ANTs) versus temperature.

Many other compounds exhibit a considerable increase in maximum hardness values in solid solutions as compared with the constituent elements or even the individual compounds themselves. These maximum hardness values of many compounds are employed in developing hard and super-hard tools, as well as heat-resistant and other materials. The questions of successive increase in hardness with increasing number of components in metallide solid solutions, the reaching of solubility limits, and decomposition reactions of supersaturated solid solutions of these metallides are very interesting. All this opens up new, hitherto little-used possibilities for developing materials harder than those existing today.

Important results were obtained in a comparative study of the heat resistance of compounds, pure metals, and solid solutions of the metals and metallides from which they are formed. We reported distinctive properties in a number of cases in [220]. Thus in the case of the metallide Ni_3Al and alloys based on it [220-221], it was established that this compound has higher heat resistance than nickel and its solid solutions with aluminum [221]. Investigation of several alloys based on Ni_3Al, containing elements having limited solubility in it, showed the possibility of developing an alloy with higher heat resistance than many nickel-based heat-resistant alloys.

The high properties of such alloys are confirmed by the change in yield point σ_{02} of nickel and a number of heat-resistant nickel alloys with respect to temperature. The corresponding curves for nickel and four heat-resistant nickel alloys including the Ni_3Al-based alloy denoted by ANTs are shown in Fig. 92. As is evident from the curves, the yield point of this alloy at high temperatures (800-1100°) is higher by 20-40 kg/mm² than those of nickel and ordinary heat-resistant nickel alloys.

In the case of continuous solid solutions of the metallide system Ti_6Al-Ti_3Sn, which we studied, alloys of this section also were found to be considerably strengthened. This made it possible to develop titanium alloys based on metallides, which were more heat-resistant than those based on titanium solid solutions [140, 141]. As the curves given in Fig. 93 show, the new heat-resistant titanium alloys, based on solid solutions of the metallides Ti_6Al and Ti_3Sn, have the same long-term tensile strength at 700° as heat-resistant austenitic steels. In this case titanium alloys based on these compounds have slightly more than half the specific gravity of heat-resistant nickel-based alloys.

In our investigations of the relation between the heat of formation (dissociation) and the heat resistance of $MeNi_3$-type metallides [239], it was shown that in the series of compounds $FeNi_3$, $MnNi_3$, $CrNi_3$, VNi_3, and $TiNi_3$, the strength of chemical bonding increases in that order from $FeNi_3$ to $TiNi_3$ and is related to their heats of formation.

The heat resistance of these compounds, whether they are formed from solid solutions or on crystallization, depends on the strength of chemical bonding and, in conformity with their heat of formation, increases in the order $FeNi_3 \rightarrow MnNi_3 \rightarrow CrNi_3 \rightarrow VNi_3 \rightarrow TiNi_3$. These compounds have higher heat resistance than the solid solutions from which they are formed (except the metallide $TiNi_3$, which is formed on crystallization).

Examples of such strengthening of metals as a result of dispersive separation of excess phases from supersaturated solid solutions are given in a monograph on the physicochemical bases of heat resistance in alloys [220].

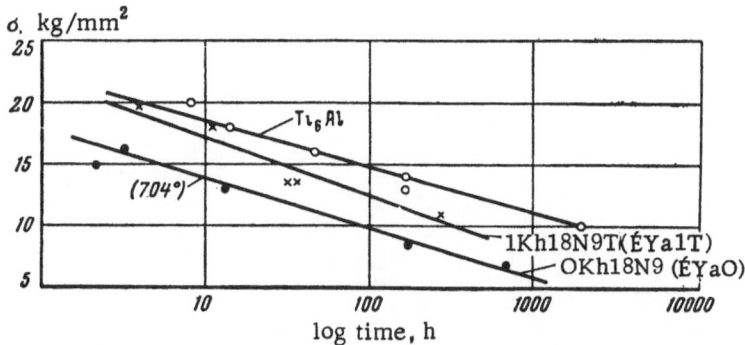

Fig. 93. Long-term tensile strength at 700° of titanium alloy based on compound Ti_6Al and heat-resistant austenitic steels versus stress.

Metallides and their solid solutions play a major role in enhancing the practically important properties of metallic alloys based on simple and many-component systems. Numerous modern technical alloys based on aluminum, iron, nickel, etc., have many components. In order to provide specified high properties (mechanical, physical, etc.), alloying elements are used which, as a rule, form limited solid solutions and compounds with the base metal. Owing to the presence of many components, these alloys contain not only the main solid solutions, but also phases based on one or several compounds. Among the latter, binary and ternary compounds may be encountered in the form of individual structural constituents—metallide solid solutions, eutectic mixtures, or finely dispersed phases formed by decomposition of supersaturated solid solutions in the process of aging. All these structural constituents in the alloys increase considerably the hardness, tensile strength, and heat resistance and improve various other physical properties. Finely dispersed phases of metallides, formed under certain conditions of transformation in the solid state [16, 17, 220], are particularly important in this respect. This is proved in many instances of high-strength aluminum and magnesium alloys, as well as iron- and nickel-based ones. Many interesting data in this connection have been obtained in the case of complex alloys of titanium, molybdenum, niobium, and other refractory metals.

CHAPTER VIII

APPLICATIONS OF METALLIDES AND ALLOYS
OF METALLIDE SYSTEMS

Metallides and alloys based on them, as was shown above, have special mechanical, physicochemical, and other properties. In this respect they are far superior to pure metals and their alloys. Many practically important properties of metallic materials are achieved only by metallides and materials based on them. Striking examples of this are the high values of high-temperature oxidation resistance and heat resistance of materials based on borides, carbides, silicides, and nitrides of refractory metals. Such compounds of titanium, niobium, molybdenum, tungsten, etc., can be used at temperatures up to 1500-2000° or more, whereas the metals forming these compounds are quite readily oxidized above 500-600°.

Important semiconductor properties — high values of forbidden-gap width, electrical resistivity, thermal resistivity, etc. — are achieved only in semiconductor compounds and their solid solutions.

In considering superconductor materials it was noted that very few superconductors are pure metals. Moreover, they have relatively low superconductivity characteristics (critical transition temperature, critical magnetic field, critical current density, etc.). Without the discovery of new metallides with especially high superconductivity parameters, one could not imagine those real possibilities for development of superconductor technology which have opened up before us in connection with the outstanding superconductor properties of Nb_3Al, Nb_3Sn, V_3Si, and V_3Ga.

Thus it can be said that the design of many new devices and instruments was based on recently discovered metallides with high magnetic and emission properties; in this case the special properties of these materials played an indispensable role. As may be inferred from these numerous examples, metallides already have found practical application; they open up new prospects for the development of technology.

The history of the discovery and application of metallides and alloys based on them is interesting. As we stated above, systematic investigations of metallides were begun in connection with the development of physicochemical analysis. This refers approximately to the beginning of this century and was connected directly with the development of Kurnakov's fundamental ideas in the field of general and inorganic chemistry.

In the second decade of this century he and his students investigated many metallic systems and discovered dozens of new metallides in them. In the initial stage, however, this work was limited to theoretical investigations in the field of chemical interactions of metals and studies of equilibrium diagrams of metallic systems. In this way they discovered a number of magnesium compounds with Group IV elements — Mg_2Pb and Mg_2Sn (which are called magnesium plumbide and stannide) — and Group V elements — Mg_2Bi_3 and Mg_2Sb_3 (called magnesium bismuthide and antimonide).

During these years many compounds of metals in Groups I, II, and III, subgroup A, with elements of Groups I-VI, subgroup B, were discovered; they are called aurides, argentides, cadmides, mercurides, aluminides, arsenides, selenides, etc. Many of these metallides, as subsequent investigations showed, are semiconductors.

Investigations of the equilibrium diagrams of Group VIII metals (Ni, Co, Pt, Pd) with bismuth, also started at the beginning of this century, led to the discovery of several nickel, cobalt, palladium, and platinum bismuthides which later proved to be superconductor compounds. In those years, however, these investigations bore the character of general "academic" work, directed toward determining the compositions, structures, and certain properties of these compounds in connection with the general study of reactions between metals and equilibrium diagrams.

Fifty years ago no one had yet studied the semiconductor and superconductor properties of metallides or imagined the possibilities which have opened up in the last 15-20 years in the field of application of metallides having semiconductor, superconductor, magnetic, and other properties. The very concept and word "semiconductor" appeared in science and engineering only 20-30 years ago. Compounds capable of acting as semiconductors were discovered 50-60 years ago. The situation is about the same with regard to superconductor compounds. Despite the fact that the compositions and structures of a number of metallides had already been determined several decades earlier, the "second birth" of these metallides and their practical application as superconductor compounds have taken place only in the last 15-20 years.

Only during these last years, in connection with unprecedented technical progress, as well as the successful development of physics and chemistry and new methods for investigating the fine structure of matter, the nature of chemical bonding, and the effect of these last on a number of mechanical and physicochemical properties of materials, has it become possible to study metallides systematically and determine their fields of practical application. The introduction of the varied class of metallides into technology demonstrates conclusively the important role of theoretical investigations in the field of metal physics and metal chemistry, the very close relation between these scientific disciplines and practice, and the fact that they have become a productive force. Many new branches of engineering have been and are being created, based mainly on recently discovered metallides with special physical properties. As noted above, semiconductor, superconductor, and other compounds may be taken as examples.

Today the fields of application of metallides in various branches of engineering have grown to such an extent that it would be difficult to review them more or less fully. For this, one would have to write a separate book or special technical handbook on applications of metallides, describing a variety of devices, instruments, and equipment, the development and production of which have become possible only since the new metallides were discovered and put into production. Data given in the special literature [219, 227, 228, 240, etc.] convey only a partial idea of all this.

In connection with the purposes of this monograph we limit ourselves to certain instances of the use of metallides and alloys based on them as materials in various technical fields. Collected literature data on fields of application of metallides and alloys based on them are given below, and indications are given of their compositions, some specific properties, and fields of application.

Data are arranged in the following order:

1) refractory, heat-resistant, and high-temperature oxidation-resistant metallides;
2) metallides with magnetic properties;
3) metallides with superconductor properties;
4) metallides with semiconductor properties;
5) metallides for cathodes in electronic technology;
6) metallides for nuclear power engineering;
7) metallides for catalysts;
8) chemically resistant metallides in aggressive media;
9) metallides for quantum generators — lasers.

These data could be classified differently, in particular, by groups of superhard, wear-resistant metallide compounds, high-strength compounds in structural steels and alloys, metallides with special mechanical, electrical, thermal, and other properties, etc. However, the volume of the present book does not permit this.

Collected Data on the Properties and Fields of Application of Some Metallides

and Alloys Based on Them

1. Heat-Resistant Metallides [217-221]

Ni_3Al and alloys based on it, containing Mo, W, Cr, etc.	M. p. of Ni_3Al 1385° (formed in peritectic reaction); high yield point and short-period creep limit at temperatures up to 1100°. Oxidation resistance up to 1100-1200°. Used for casting heat-resistant and high-temperature oxidation-resistant articles.
NiAl and alloys based on it, containing Mo, TiC, $MoAl_2$.	M. p. of NiAl 1650°, deformable at temperatures above 1300°; heat-resistant and high-temperature oxidation-resistant material; brittle at room temperature. Used for castings with working temperature up to 1200-1300°.
$ZrNi_4$ and possible alloys based on it.	M. p. of $ZrNi_4$ 1600°; deformable when cold; no detailed information on other properties. Interesting as a metallide, plastic at low temperatures.
TiNi and possible alloys based on it.	M. p. of TiNi 1310°; deformable when cold; has high tensile strength up to 600-800°; interesting as a plastic, heat-resistant metallide with lower specific gravity than nickel alloys, and also as a basis for preparing plastic alloys.
$TiCr_2$ and alloys based on it.	Kurnakov compound. Temperature of formation of $TiCr_2$ 1350° from β-solid solution of titanium; high modulus of rupture and long-term tensile strength at 1000-1100°; brittle at room temperature. Used as heat-resistant material for castings.
Ti_6Al and alloys based on it, containing Zr, Sn, Ti_3Sn, etc.	Formed from α-solid solution of Ti; retains tensile strength up to 600-700°; has higher limit of creep up to 600° than titanium alloys; is a deformable, heat-resistant material up to 600-700° with density 4.5 g/cm^3. Proposed as a new, heat-resistant titanium alloy [140-141].
Ti_3Al and alloys based on it, containing Zr, Sn, TiSn, Ti_3Al—Ti_3Sn solid solutions	Formed from α-solid solutions of Ti; retains high tensile strength and limit of creep up to 700-800°; deformable in the hot state; diminished plasticity at room temperature. Has high elastic limit, about 1.5 times that of pure titanium [140, 141].
Ti_2Al (hypothetical) and alloys based on it, containing V, Zr, Sn, etc.	Formed from α-solid solution of Ti; strength retained up to 900-1000°; has high high-temperature oxidation resistance and heat resistance; brittle at room temperature. Proposed for castings with heat resistance up to 800-900° [140].
TiAl and alloys based on it, containing Nb, Ni, Zr, NiAl, etc.	Temperature of peritectic formation reaction 1460°; high heat resistance and oxidation resistance up to 1100-1200°; brittle at room temperature; strength increases on heating to 900-1000°. Proposed as heat-resistant casting alloy [140].

TiB_2 and alloys based on it, containing Ti, CrB_2, etc.

M. p. of TiB_2 2980°; high hardness and modulus of elasticity; long-term tensile strength up to 1200°; heat-resistant metal—ceramic material. Used for preparing thermocouple sheaths, crucibles, and linings for melting aluminum and other light metals [53, 77, 78].

ZrB_2 and alloys based on it (borolites) with binders consisting of Mo, Cr, and other borides of V, Ni, Cr, Mo, W.

M. p. of ZrB_2 3250°; high high-temperature oxidation resistance, heat resistance, and heat stability. Used for preparing high-temperature thermocouple sheaths, resistance-furnace heaters, melting crucibles, tubes for transfer of molten metals, etc. [54, 77, 78].

TiC and alloys based on it (cermets), containing WC, Cr_3C_2, Co, Ni, solid solutions with TaC, NbC, TiB_2, etc.

M. p. of TiC 3147°; high hardness and modulus of elasticity; heat resistance and oxidation resistance up to 1100-1200°. Used in the form of cermets for preparing small components of high-temperature devices, gas turbine blades, and other purposes [54, 77, 78]

ZrC and alloys based on it.

M. p. of ZrC 3530°; high heat resistance and high-temperature oxidation resistance, low neutron-absorption cross section. Used as constituents of heat-resistant materials in nuclear power engineering and for making crucibles, boats, and tubes which are stable in molten metals [54, 77, 78].

$MoSi_2$ and alloys based on it, containing Cr, Al, Si, and WSi_2.

M. p. of $MoSi_2$ 2030°; high oxidation resistance up to 1500-1700°. Used for components of gas turbines, rocket-motor combustion chambers, rockets, thermocouple sheaths, resistance-furnace heaters for temperatures up to 1500-1600°, working in air and oxidizing media [55, 77, 78].

AlN (aluminum nitride).

M. p. 2400°; high stability on fusion of aluminum alloys; fire-resistant material. Has recently received attention as a semiconductor material [77].

TiN and alloys based on it; solid solutions containing nitrides and oxides.

M. p. 3205°; high heat resistance and oxidation resistance up to 1000-1100°. Used for protective coatings on titanium and in other fields [54, 77, 78].

Transition metal beryllides $TiBe_{12}$, $ZrBe_{12}$, $MoBe_{12}$, $NbBe_{12}$, Zr_2Be_{17}, and solid solutions based on them.

High melting points (1500-2000° or higher); oxidation resistance up to 1400-1500°; they retain high tensile strength up to 1510°. New heat-resistant materials—beryllides—may be used in nuclear equipment up to working temperatures of 1200-1650° [77, 223].

2. Metallides—Magnetic Materials [226-236]

Fe_3Si, Fe_3Al, and Fe_3Si—Fe_3Al solid solutions (Alsifer).

Kurnakov compounds with ordered structure and their solid solutions; they have high magnetic permeability. Used in instruments in the form of pressed powders (owing to their brittleness).

$FeNi_3$ and alloys based on it, containing Mo, Al, Mn (Permalloy).

Kurnakov compound with ordered structure and its solid solutions, with high initial and maximum permeabilities and low coercive force. Used for making telephone apparatus, sensitive relays, etc.

MnNi$_3$ and MnNi$_3$—FeNi$_3$ solid solutions (Megaperm industrial alloys).

Kurnakov compounds with ordered structure; has high saturation induction; in solid solutions of compounds the latter reaches values exceeding those for the pure compounds MnNi$_3$ and FeNi$_3$.

Fe$_2$Co, FeCo, and alloys based on them.

Kurnakov compounds with ordered structure; have high initial and maximum permeabilities. Used in magnetic circuits of electromagnets and permanent magnets.

FePt, FePd, FePd$_3$, CoPt, and alloys based on them.

Kurnakov compounds with ordered structure; under optimum conditions of heat treatment they show high saturation induction and coercive force.

Alloys near the compounds Fe$_2$NiAl, containing Co and Cu (alloys Mishima, Alnico, etc.).

Alloys for permanent magnets with high magnetic characteristics.

Fe$_2$B, FeB$_2$, Fe$_2$Ge, Fe$_3$P, Co$_5$As$_2$, Co$_2$B, Co$_2$P.

All compounds have ordered structure and are ferromagnetic. They are important as ferromagnetic materials.

Alloys near the compounds Cu$_2$MnAl, Cu$_2$MnSn, Cu$_2$MnGa, Cu$_2$MnZn, Cu$_2$MnIn.

Heusler alloys have ordered Fe$_3$Al-type structure; they are ferromagnetic. Alloys have high magnetic saturation and coercive force and maximum values of Curie points. The alloy Ag$_5$MnAl has especially high coercive force.

3. Superconductor Metallides [243-247]

Intermediate phases of system BiPb.

The first solenoid (1930, Netherlands) with a maximum magnetic field of 20,000 G was based on alloys of this system.

Alloys of system Mo—Re, having composition MoRe$_3$.

Critical temperature 9.26° K; cubic, α-Mn-type structure. Solenoid with maximum magnetic field of 20,000 G was developed (1960, USA).

Alloys of system Nb—Zr, corresponding to intermediate two-phase region (25-35% Zr).

Alloys of this system, contrary to metallides, are plastic; cold-drawn wire may be obtained from them. Solenoids having magnetic fields of 30,000-40,000 G were made (1954) in USA and USSR.

Alloy of the system Nb—Sn, corresponding to the compound Nb$_3$Sn, and solid solutions based on it.

T_C = 18.05° K, β-W-type cubic structure. Solenoid with high magnetic field (80,000-100,000 G) and high current density (over 10^5 A/cm^2 at 88,000 G). Some indirect experiments indicate that the magnetic field may rise as high as 200,000 G.

Alloy of the system V—Ga, corresponding to the compound V$_3$Ga, and solid solutions based on it.

T_C = 16.5° K, β-W-type cubic structure. Of all known superconductor compounds, V$_3$Ga enables one to reach the highest magnetic field intensity, equal to 400,000 G at a current density above 10^5 A/cm^2.

Other superconductor compounds of compositions V$_3$Si, Nb$_3$Al, Nb$_3$Ga, Ta$_3$Sn, NbC, NbN, TiN, and solid solutions of these compounds.

May be used as materials for solenoids with magnetic fields of 100,000 G or more.

Metallides for Nuclear Engineering [219, 248, 249]
(other than heat-resistant structural materials)

UC, UC_2, UC_2 + graphite, UN, U_3Si, U_3Si_2, ThC, UAl_4, UAl_3, UFe_2, UBe_{13}. Solid solutions based on metallide systems UC — ThC, UC — ZrC, etc.

Used in the form of dispersed mixtures for heat-evolving elements of atomic reactors; metallides must be compatible with structural materials and have minimal values of neutron-capture cross section, as well as high thermal conductivity, melting point, and other properties.

Be_2C_6 (with a Pt or ceramic antioxidation coating), hydrides of some transition metals: TiH, ZrH, TiH_2, ZrH_2, CeH, NbH, VH_2, etc.

Materials for slowing fission neutrons in atomic reactors. The listed materials—beryllium carbide, hydrides of some transition metals — are promising as moderators and are so used.

Some hafnium metallides, materials based on boron carbide (B_4C) (borals), SmB_6, EuB_6, GdB_6, intermediate silver—cadmium, finely dispersed inclusions of the isotope B^{10} in TiB_2, Zr^n_2, B_4C, Zircalloy, stainless steel.

Materials for reactor control rods, radiation shielding, neutron moderators and absorbers, and γ-ray attenuators (absorbers).

5. Semiconductor Metallides [97, 100, 105, 240-251]

PbS, PbSe, PbTe, InSb, Tl_2S, CdS, CdSe, CdTe, Bi_2S_3, and possibly solid solutions and alloys based on them.

These compounds are semiconductors with predominantly ionic bonding, having a narrow forbidden gap; they are appreciably photoconductive in the infrared. Used in infrared radiation detectors, photocells, photoresistors, transistors, and for other purposes. The most sensitive photoresistor materials are CdS and PbS.

Oxide semiconductor materials, with mainly or wholly ionic bonding: Cu_2O, NiO, CoO, MgO, TiO, TiO_2, and in the form of solid solutions or mechanical mixtures of oxides with SiC (as a varistor).

The temperature coefficients of electrical resistivity of these compounds are many times those of the pure metals, and are negative. These materials are used as thermistors, varistors, and other nonlinear resistors, and also for precise temperature measurement (precision within 0.005°), time relays, inertialess resistors, vacuum gauges, and stress stabilizers.

ZnS, CdS, CaS, SrS, and BaS with additions of Cu, Mn, Sn, Se (for long-duration luminescence—luminophors), ZnS with additions of Ag, Ni, Co, Cd (for short-duration luminescence—fluorescence), etc.

These compounds are very important for obtaining "cold light" by emission from semiconductor luminophors. Used in fluorescent lamps, for constructing particle counters in nuclear physics, fluorescent screens in x-ray apparatus and electron microscopes, fluorescent paints, devices for night vision.

GaAs, InSb, InP, InAs, AlSb, GaP, and possibly solid solutions based on them.

Used as high-temperature rectifiers, diodes with variable capacitance, microwave detectors. Diodes made of compounds providing high-temperature rectification and a broad forbidden gap have considerable advantages over those made of germanium.

HgSe, HgTe, InAs, InSb, PbTe, and solid solutions based on them.

All these compounds have high electron and hole mobilities. The large Hall emfs of these semiconductors enable one to use them in constructing Hall-effect detectors and for measuring current intensity and direction. From them are made devices—Hall-effect detectors—which may be used in current transformers, protective relays, for measuring magnetic fields, and other purposes.

Bi_2Te_3, Bi_2Se_3, GaP, GaAs, InP, AlSb, Bi_2Te_3—Bi_2Se_3, PbS, PbTe, InSb, MnTe, CeS, ZnSb, CdSe, CdTe, CdS, and solid solutions based on them, having lower thermal conductivity than the pure compounds.

They have thermo-emf and relatively high electrical conductivity with comparatively low thermal conductivity. Used in the form of single crystals, polycrystalline masses, and thin films as materials for thermoelectric generators, thermoelectric refrigerators, and converters of solar energy into electrical [solar batteries and other high-efficiency cells (15-25%)]. Working temperatures 600-650°, which can be raised to 1000-1030° [251].

AlSb, GaAs, InP, Cu_2O, CuS, SiC.

Semiconductors having p-n junctions. Used for preparing p-n-type semiconductor rectifiers and p-n-p-type transistors. Refractory semiconductors (GaAs, GaP, SiC) are especially important.

AlN, AlP, BN, etc.

Semiconductors with broad forbidden gap, retaining semiconductor properties up to high temperatures.

6. Metallides for Cathodes in Electronics [54, 219]

ZrC, UC, NbC, and solid solutions ZrC—UC, UC—HfC, UC—NbC, UC—TaC.

Materials for emitters at working temperatures 2000-2600°, with maximum current density in converters of heat into electrical energy.

CoB_6, SrB_6, BaB_6, and solid solutions SrB_6—BaB_6, YB_6, LaB_6, CeB_6.

High thermoemissive properties, low specific gravity, used in compositions of cathodes in electronics.

7. Metallides for Catalysts [53, 55, 219]

Mo_2C, WC, Fe_2C, $MoSi_2$, rare-earth metal sulfides, borides of iron and other metals, suboxides of certain metals, semiconductor compounds.

Used in the chemical industry for catalytic processes, dehydration of alcohols, oxidation, hydrogen addition, and other processes.

8. Chemically Stable Metallides in Aggresive Media [219]

Cr_3C_2 and alloys Cr_3C_2—TiC—Ni, Cr_3C_2—WC—Ni.

Chemically stable toward various mineral and organic acids in the cold and hot states. Metal—ceramic materials based on the carbide and chromium have high acid resistance, mechanical strength of about 65 kg/mm^2, and low coefficient of thermal expansion ($6.4 \cdot 10^{-6}$). Used for preparing filters in the chemical industry.

Metallide solid solutions $CeAg_3$—$LaAg_3$.

High corrosion resistance in water and in a current of oxygen, compared with that of the pure compounds [126].

TiB_2, ZrB_2, MoB_2, their solid solutions and alloys with iron-group metals.

High corrosion resistance in concentrated phosphoric acid (up to 105% H_3PO_4) at temperatures up to 200°. The alloys TiB_2 + 19% Ni and ZrB_2 + 12% Ni are most resistant in molten fluorides and in lithium on heating to 1000-1100°.

9. Metallides for Quantum Generators [252].

GaAs, GaP, InAs, InP, and solid solutions of the systems GaAs—GaP and GaP—InAs.

GaAs gives light of wavelength 6500 A; GaP—8400 A, InP— 9100 A; InAs—91,000 A. GaAs—GaP solid solutions give light of wavelengths from 6500 to 8400 A; GaAs—InAs solid solutions, from 6500 to 34,000 A. All these compounds and their solid solutions are used as materials in recently developed high-efficiency quantum generators (Scientific American, 209(1):34, July 1963).

The above brief list of properties of various types of metallides shows the extent of possible application of these materials in many fields of machine building and construction of instruments and apparatus. Presumably, further investigations of metallides, especially their properties and interactions, and synthesis of new compounds will lead to the discovery of new fields of application for them. Extensive potential possibilities for the practical use of metallides and alloys based on them are latent in these investigations.

CHAPTER IX

SOME QUESTIONS OF METALLIDE CHEMISTRY

Investigations of reactions between metals are important in the general development of inorganic and metal chemistry. As noted above, our consideration of theoretical and experimental data on metallide formation is based on the periodic law of chemical elements. Application of this reveals the main rules governing interactions of metals, the appearance of solid solutions in some metallic systems and compounds in others, and the occurrence of compounds with metallic bonding in some systems and covalent, ionic, or mixed chemical bonding in others.

In connection with this, questions of the disposition of metals in the periodic system are important for discussing certain aspects of reactions between metals. The similarity of atomic structure in some elements and dissimilarity in others follow from this. The principles of division of the elements into metals and nonmetals according to the electronic structure of the atoms, customary in the chemical literature, enable one to separate metals from nonmetals more or less arbitrarily.

In this book, without discussing the expediency of the arrangement of elements in Mendeleev's periodic table, we have presented it in slightly altered form (Fig. 94). In this form it is easy to show the boundary between metals and nonmetals. This boundary is shown in Fig. 94 by a diagonal dashed line; elements to the left of this line, beginning with Be, Al, Ge, Sb, and Po, are metals, whereas those to the right, beginning with B, Si, As, Te, and At, are nonmetals. Of them, B and Si are typical nonmetals, whereas the heavier elements of this series — As, Te, and At — acquire certain metallic properties as the atomic number increases. In chemistry they are, with some justification, called semimetals.

Using this extended scheme (see Figs. 1 and 2 above), we presented the following metallochemical properties of the elements: atomic radii, electronegativities, and ionization potentials. This method of representing the properties of the elements graphically by groups, rather than by periods, as was done earlier, brings out more vividly the rules of variation of these properties with position in the periodic system. Moreover, this order of arrangement and juxtaposition of the chosen properties of the elements provides some data for determining the subgroups in which they fit best. The places of B and Al, as well as Be and Mg, in the periodic system were not clear until recently. Some authors place boron and aluminum in Group IIIA, in the column of scandium, yttrium, etc., whereas others refer them to Group IIIB, in the column of gallium, indium, and thallium. The positions of beryllium and magnesium also are rather uncertain: Some refer them to Group IIA, and others, to Group IIB. According to the metallochemical properties given above (see Figs. 1 and 2), boron and aluminum certainly fit better in Group IIIB. With regard to beryllium, also, one may say that it belongs to Group IIB, whereas the position of magnesium is less certain, although it too can be placed, according to several metallochemical properties, in Group IIB.

The order in which we arranged these elements in the periodic system is confirmed by several metallochemical properties of boron and aluminum, as well as beryllium and magnesium. As is evident from Fig. 1, placing B and Be in Groups IIB and IIIB results in an orderly sequence (almost linear dependence) of decrease in atomic radius of the following series of elements in the second period: Be → B → C → N → O. The variation of ionization potential in this series is less orderly; nevertheless, these elements (except beryllium) have the highest ionization potentials in their respective groups, increasing in the order B → C → N (see Fig. 2).

The large atomic radii and low ionization potentials of elements on the left-hand side of the table (Groups I and II excluding beryllium), on the one hand, and the smaller atomic radii and higher ionization potentials of elements on the right-hand side (Groups IV-VI), on the other, determine the characteristics of reactions of metals with other elements.

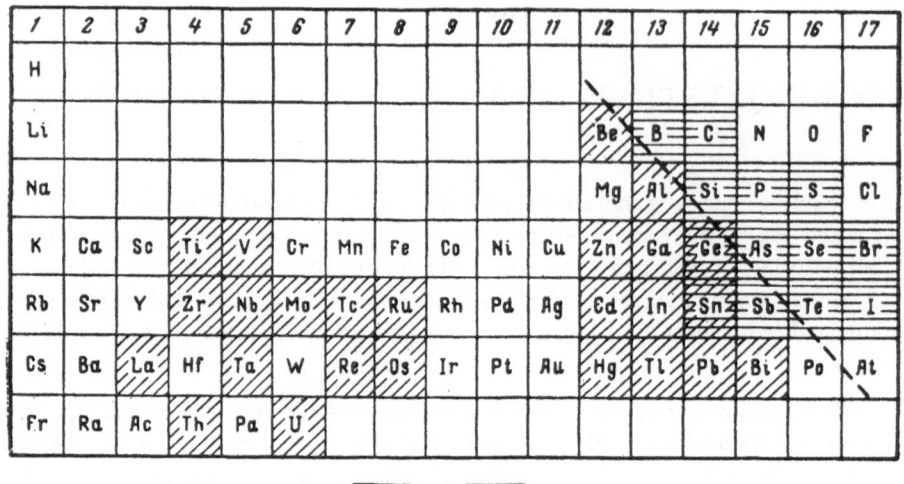

Fig. 94. Periodic system and distribution of superconductor and semiconductor elements in it: 1) semiconductor elements; 2) superconductor elements.

These peculiarities in the chemical behavior of metals are also favored by the analogous order of variation in electronegativity of elements within each group. These aspects of reactions of metals will be considered below from a general viewpoint.

Further grounds for placing Be, B, and Al in Groups IIB and IIIB consist in the fact that all these three highly electronegative elements tend to give many compounds with metals. In this case, participating as electronegative elements, they determine the names of the compounds—beryllides, borides, and aluminides. Many of these compounds were described above. As regards magnesium, its role as an electronegative element in metallide formation is limited, and accordingly we rarely encounter such names as magneside.

The extended form of representation also enables one to distinguish readily two families of elements in the system, which have special physical properties not possessed by other elements. The first and second families consist of superconductor and semiconductor elements, respectively. The first family comprises 24 elements, all of which are superconductor metals. They are shaded by diagonal lines in Fig. 94. Their number and position in the periodic system are less definite; they encompass a large group of metals of the fourth to seventh periods, excluding certain metals of Groups IV (Hf), V (Pa), VI (Cr and W), VII (Mn), and VIII (ferromagnetic metals Fe, Co, Ni), as well as platinum metals (Rh and Pd, Ir and Pt). This family does not include metals of Group IB (Cu, Ag, and Au); Mg is excluded from Group II, B from Group III, the nonmetals C and Si from Group IV, and N, P, As, and Sb from Group V.

It must be admitted that the small number of superconductor elements revealed by modern data does not permit one to formulate well-grounded rules of their arrangement in the periodic system, and their number obviously is still incomplete. This property is very sensitive to impurities. Further, more thorough investigation of this question, using exceedingly pure metals, will possibly enable one to discover other metals with superconductor properties, to arrange them in a more orderly way in the periodic system, and to formulate some general rules governing this arrangement.

The number of semiconductor elements is even more limited. In Fig. 94 these elements are shaded by horizontal lines; there are 12 of them in all. Here it is characteristic that this family of elements is grouped on the right-hand side of the table and includes both metals and nonmetals. Two elements — Ge and α-Sn — have superconductor and semiconductor properties simultaneously. The semiconductor metals include elements located at the boundary between metals and nonmetals: Ge and As on one side and Sb and Te on the other. The rest of the elements are nonmetals: Group III — B, Group IV —C and Si, Group V —P and As, Group VI—S and Se, and Group VII—I.

These elements differ considerably among themselves in the main characteristic property of semiconductors — forbidden-gap width. This increases with both the electronegativity and degree of nonmetallic character of the element. As is evident from Table 15, the smallest forbidden-gap widths occur in α-Sn (0.08 eV) and Sb (0.12 eV), and the largest in S (2.5 eV) and carbon (diamond 6 eV). The paucity of superconductor and semiconductor elements considerably limited their use. Moreover, they are the basis for the formation of a large number of metallides having superconductor or semiconductor properties. One or both components in most of these compounds are superconductor or semiconductor elements.

As noted above, among superconductor compounds there are cases where such metallides are formed even though no superconductor elements are present. The existence of superconductor compounds containing nonsuperconductor metals (tungsten, hafnium, radium, and iridium) gives rise to certain hypotheses regarding the possible presence of superconductivity in these metals. It also attests that the superconductor metals in the periodic system have not all been identified as yet. Among semiconductor compounds there are apparently none formed by the combination of two nonsemiconductor elements. All the semiconductor compounds considered above must contain one or two semiconductor elements; these compounds, depending on the difference in chemical properties between the component elements, have a wide range of types of chemical bonding. They may vary from purely covalent to purely ionic through a number of intermediate types. This is indicated by the forbidden-gap widths ΔE of these compounds (see Tables 15 and 37).

As is evident from these data, the variation of this property of AB- or A_2B_3-type semiconductor compounds is governed by the following rule: The forbidden-gap width increases with the difference in metallochemical properties of the reacting elements. This is evident in Fig. 15 for compounds of Al, Ga, and In with P, As, and Sb. It can be shown also in the form of the following series for zinc and antimony sulfides, selenides, and tellurides:

Compound	ZnS	ZnSe	ZnTe	Sb_2S_3	Sb_2Se_3	Sb_2Te_3
ΔE, eV	3.6	2.68	2.25	1.70	1.2	0.35

It may be concluded from this that the forbidden-gap width for these compounds increases with the difference in metallochemical properties (see Table 1 and Figs. 1 and 2) between elements A and B (between Zn and S, Se, or Te, and between Sb and S, Se, or Te).

The fact that ΔE is 1.70 eV in Sb_2S_3, where the difference in properties of Sb and S is large, whereas it is only 0.35 eV in Sb_2Te_3, where Sb and Te are adjacent in the periodic system, is characteristic in this respect. The same rule may be demonstrated in cases of other semiconductor compounds, proceeding from the ΔE values given in Table 16. The fraction of ionic bonding decreases in the same order in this series of compounds. It is largest for ZnS and smallest for ZnTe.

From this follows the important conclusion—important for the present study—that the smaller the difference is between such compounds with regard to forbidden-gap width and type of chemical bonding, the more they tend to form mutual metallide solid solutions.

As we are considering in this book not only particular cases of formation of compounds with superconductor and semiconductor properties, but also the formation of all types of metallides in general, we must say that all metals, regardless of their position in the periodic system, have some ability to form metallic compounds.

It can be shown that a metal of any group in the periodic system can form compounds of some sort with elements of any other group except the inert gases. Moreover, a given metal will not form compounds with all elements. One of the problems of metal chemistry is to discover and formulate general rules revealing why compounds are formed in some metallic systems and not in others. Having such rules at one's disposal, one can use them not only to establish the compositions of metallides discovered already, but also to predict possible cases of formation of such compounds in metallide systems not studied hitherto.

The location of the reacting elements in the periodic system and the resultant atomic radii, electronic structures, electronegativities, valences, and ionization potentials were taken as the main factors determining

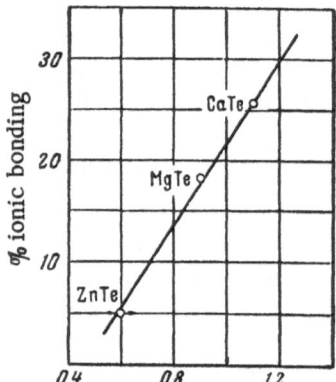

Electronegativity difference
Fig. 95. Curve of fraction of
ionic bonding in covalent compounds ZnTe, MgTe, and CaTe
versus difference in electronegativity of components.

the conditions of formation of metallides. Reactions of formation of solid solutions and metal compounds with different types of chemical bonding are considered from the viewpoint of these properties, which we regard as metallochemical properties of the elements. Depending on the nature and structure of their component elements, as well as their atomic ratios, metal compounds have different types of chemical bonding. The latter vary from purely metallic to covalent and ionic. Moreover, intermediate types of bonding are encountered among such compounds: metallic combined with covalent, covalent with ionic, and metallic with ionic.

It is easy to show that compounds with purely metallic bonding result from reactions between typical metals with nearly identical metallochemical properties. Covalent compounds are formed in reactions between metals, semimetals, and nonmetals when the difference in electronegativity between the reacting elements is roughly equal to 2 or less; when elements react whose electronegativity difference is larger than 2, compounds with purely ionic bonding are formed as a rule. It will be noted that in compounds with metallic bonding, a fraction of covalent bonding may be found, whereas compounds with covalent bonding have a certain fraction of ionic bonding. It is minimal for compounds with a small electronegativity difference between the components and increases with the latter. These questions are considered in detail for many compounds in [30]. Recent data on the increase in the fraction of ionic bonding in the series of covalent compounds ZnTe, MgTe, and CaTe (Fig. 95) [252] are a striking example. In this series, which has the common anion Te, the electropositivity of the atoms increases in the order Zn → Mg → Ca, as does accordingly the fraction of ionic bonding.

The variation in type of chemical bonding in metal compounds is affected by both the relative difference in metallochemical properties of the elements and the relative concentration of electronegative elements in the compound. This was shown above in the instances of several compounds of metals with silicon, phosphorus, sulfur, etc. Such compounds with a large number of electropositive-metal atoms have metallic bonding, and with increasing number of atoms of the electronegative element in the series of compounds $A_3B-A_2B-AB-AB_2$, the type of bonding changes gradually from metallic to covalent and ionic or, in some cases, intermediate. The concentration factor in the change in type of chemical bonding in metallides is manifested in this way.

Of the different systems of classification of chemical compounds, including metal compounds, the Kurnakov classification was adopted as a basis in this book. According to Kurnakov, all metal compounds may be divided into two types — berthollides and daltonides, which are distinguished by the structures and properties characteristic of these types of compounds in property—composition diagrams. We pointed out the essential difference in the properties of such compounds above in the cases of the Kurnakov compound VFe_3 and covalent compound Mg_2Sn (see Fig. 83).

Of a wide variety of names of metal compounds, we adopted the name metallide, giving it a broader meaning than simply intermetallic compound, or intermetallide. In the general concept of metallides we encompass all metal compounds, including strictly intermetallic ones (compounds of metal and metal), compounds of metals and nonmetals, and superconductor and semiconductor compounds having metallic, covalent, or intermediate types of chemical bonding. The concept metallide excludes only metal compounds having purely ionic bonding and compounds of nonmetals with nonmetals — the so-called nonmetallic compounds (silicon carbide, boron nitride, etc.). They should be called nonmetallides.

This type of division of metal compounds is more general than any proposed hitherto. It is also arbitrary to some extent, and the boundary line for such compounds is less definite than, e.g., between inorganic and organic compounds. Nevertheless, such a separation of metallides from the aggregate of inorganic compounds to form a large, independent group is methodologically correct. It is justified by the existence of general rules governing the interactions of these compounds.

TABLE 40 Compositions and Names of Sodium and Magnesium Metallides

Sodium metallides		Magnesium metallides	
Formula	Name	Formula	Name
With Group IVB elements			
NaSi	Sodium silicide	Mg_2Si	Magnesium silicide
NaGe	Sodium germanide	Mg_2Ge	Magnesium germanide
NaSn	Sodium stannide	Mg_2Sn	Magnesium stannide
NaPb	Sodium plumbide	Mg_2Pb	Magnesium plumbide
With Group VB elements			
Na_3As	Sodium arsenide	Mg_3P_2	Magnesium phosphide
Na_3Sb	Sodium antimonide	Mg_3As_2	Magnesium arsenide
Na_3Bi	Sodium bismuthide	Mg_3Sb_2	Magnesium antimonide
		Mg_3Bi_2	Magnesium bismuthide

In naming metallides on the basis of their component metals and in writing their formulas, we adopted the order recommended by the Commission for the Nomenclature of Chemical Compounds, Academy of Sciences, USSR [36]. In accordance with this recommendation, the electropositive metal is written first in the formula, followed by electronegative elements in the order of increasing electronegativity (see Table 1).

Following the Commission's recommendation (see above), one obtains such names as nickel aluminide, magnesium antimonide, etc. It will be noted that a given metal, according to its position in the electronegativity series and participation in metallide formation, may appear in two places in names of compounds: For instance, the metallide Ni_3Al is nickel aluminide, whereas the metallide AlSb is aluminum antimonide. In the first compound Al is an electronegative element relative to nickel, whereas in the second it is an electropositive metal relative to Sb. The use of such names for metallides in books and papers on inorganic chemistry, metallurgy, and metal chemistry will bring uniformity and order to their classification.

This order of writing formulas and names of compounds which we call metallides is exemplified in Table 40 by certain compounds of electropositive metals, e.g., those of Groups I and II (Na and Mg), with electronegative elements, e.g., those of Groups IVB and VB.

The list of metallides with such names could be extended, particularly for the large class of semiconductor compounds $A_{II}B_V$, $A_{III}B_{VI}$, etc.

This principle of names may be used also for compounds formed by any two metals with nearly identical electronegativity values. For instance, the compounds $AuCu_3$ and AuCu, formed from solid solutions of the system Cu—Au, should be called gold cuprides and not the reverse—copper aurides—since gold is more electropositive than copper in the electronegativity series. Also, one should call chromium—cobalt compounds (CrCo, σ-phase) chromium cobaltides and $TiCr_2$ titanium chromide, since Co is electronegative relative to Cr, as is Cr relative to Ti. As regards the atomic ratios in the compositions of metallic compounds, it is well known that metallides, as a rule, are not subject to valence relations, and their compositions may vary from the simple ratios A_3B, A_2B, AB_2, and AB_3 to complex ones such as, e.g., AB_5, A_6B_7, A_6B_{23}, AB_{12}, AB_{17}, etc.

These atomic ratios in the compositions of metallides have not as yet received proper scientific explanation. It should be noted merely that the most electronegative metals (Al, Be, etc.) have a pronounced tendency to form compounds in which their atoms are present in large numbers ($CrAl_7$, $TiBe_{22}$, $NbBe_{17}$, etc.). These questions also require independent discussion and development.

All metallides are crystalline in the solid state. The structural relations among such compounds were worked out by numerous investigations of their crystal chemistry. It was found that most metallic compounds crystallize in the same basic structure types as pure metals. Among them are often encountered the structure types NaCl, CsCl, NiAs, ZnS, $MgCu_2$, $MgZn_2$, etc., which are well described in [58-64]. These compounds become more complex in structure with increasing difference in chemical properties between the reacting elements.

As can many inorganic compounds, metallides can interact. Such interactions result in the formation of metallide solid solutions (continuous and limited) and simple mechanical mixtures, and ternary and more complex compounds appear in many cases. Hence it is essential to study and determine the rules of interaction of metallides, equilibria between them, compositions and structures of the phases appearing in these reactions, and the functional relation between composition and various properties. This constitutes a new division of metal chemistry, the systematic study of which is very important.

Solid solutions of metallides are analogous to solid solutions of metals in many respects. Crystallochemical and x-ray structure investigations of metallide solid solutions showed that the latter, as in the case of metals, occur in two types—one formed by substitution of atoms in the structure of the solvent compound and the other by insertion of atoms in interstices in the structure.

In both substitutional and interstitial solid solutions the dissolved atoms are distributed at random in the structure of the compound. In individual cases, however, one can imagine certain phenomena of ordering, as perhaps in the formation of ternary Kurnakov compounds from solid solutions of two metallides. This was shown in [181] in the case of the system Bi_2Se_3—Bi_2S_3. In metallide solid solutions the lattice parameters vary regularly, deviating from linearity to some degree (not subject to Vegard's law). Any such deviations of the curve of lattice parameter of solid solutions versus composition from linearity attest to the manifestation of chemical affinity and the presence of interaction between dissolved atoms in the metallide solid solutions. In connection with this, however, it should be noted that the character of chemical bonding in metallides is not changed when one such compound dissolves in another to give continuous or limited solid solutions.

In both the case of substitutional solid solutions and that of interstitial ones, conditions exist determining the possibility or impossibility of total or limited mutual solubility of metallides.

As a development of theses advanced by us earlier [111-113], the main factors determining the character of interactions of metallides and the conditions of formation of solid solutions and simple mechanical mixtures between them are considered and stated more precisely in this book.

The conditions necessary for the formation of continuous metallide solid solutions are:

1) identical types of structure and nearly identical lattice constants;
2) identical types of chemical bonding in the compounds;
3) the presence in the compounds of atoms which are analogs or have nearly identical metallochemical properties;
4) identical atomic ratios.

These conditions are closely interrelated, and questions of the formation of continuous metallide solid solutions should be considered from the general viewpoint of all of them. If even one of the conditions is not met, the result will be limited continuity or the total absence of solid solutions composed of the given metallides.

The physicochemical factors listed above were confirmed and refined by many authors [114-128] in instances of formation of continuous and limited solid solutions among different types and classes of metallides. The large amount of literature and experimental material accumulated along this line made it possible to generalize, check, and prove the validity of the stated factors determining the conditions of formation of solid solutions and to use them in predicting the character of interaction of many unstudied metallide systems. Detailed analysis of these data made it possible also to uncover certain exceptions to these rules and to explain them appropriately.

The following important theses of metallide chemistry are based on a comprehensive investigation of these questions.

1. For the formation of continuous solid solutions between metallides, the latter must be isomorphous, other conditions being equal. In this respect, no exceptions have been found in the crystal chemistry of metallide solid solutions. Instances of continuous transitions by deformation, insertion, or removal in unit cells of the nickel arsenide structure type cannot be considered exceptions. This was shown in [74] in the case of gradual transition of structure types from Ni_2In to NiAs to $NiTe_2$ by removal of nickel atoms from the first structure.

2. The presence of identical types of chemical bonding together with structural isomorphism of the compounds is also an absolutely necessary condition for the formation of continuous solid solutions. The diverse character of chemical bonding between atoms and the distribution of electrons in the structure of compounds with metallic and covalent, covalent and ionic, or especially metallic and ionic bonding presupposes the impossibility of continuous substitution or insertion of foreign atoms in the structure of the solvent compound. In many cases compounds with different types of chemical bonding have different types of structure. Experimental data show that metallides with different types of bonding and different structures do not give continuous solid solutions. Some examples confirming the absence of continuity for solid solutions composed of compounds with different types of chemical bonding are given in Table 41.

Another striking example of the absence of continuity in solid solutions is the system of isomorphous metallides Mg_2Sn-Mg_2Pb, where Mg_2Sn is a semiconductor and Mg_2Pb a compound with purely metallic bonding. However, the presence of the same, covalent type of bonding in the metallides Mg_2Ge and Mg_2Si and their isostructurality determines the formation of continuous solid solutions in the system composed of them (see Fig. 51).

Accepting these theses as valid, one may use them to predict the character of interaction between metallides with various types of chemical bonding and to prove the presence or absence of continuous solubility in systems composed of them. Many examples can be given of metallide systems where continuous solid solutions are formed and where they are not formed.

3. The third condition, determining the possibility of formation of continuous solid solutions between metallides, proved to be the most complex and hence least definite. Therefore, this question must be considered in more detail.

In developing the theory of formation of metallide solid solutions, the formation of continuous solid solutions between one pair of constituent elements of the compounds was taken as one of the conditions for total mutual solubility [113]. This also is understandable, since in this case it is easy to exchange these atoms in the metallide structure without affecting its stability. This thesis has no exceptions for either substitutional or interstitial solid solutions. One can readily imagine, for instance, continuity of solid solutions in the metallide

TABLE 41. Metallide Systems in Which Continuous Solid
Solutions are Absent

Characteristics of compounds	Metallide system	
System with metallic and covalent bonding		
Structure type	Hex.	Hex.
Type of chemical bonding	Metallic	Coval.
System with covalent and ionic bonding		
	InSb—CdS	
Structure type	Cub. (ZnS)	Cub. (ZnS)
Type of chemical bonding	Coval.	Ionic

system $NbNi_3-TaNi_3$, due to total exchange of atoms of the analogs niobium and tantalum in the structure of solid solutions of these isomorphous compounds. One also can imagine exchange of nitrogen and oxygen atoms in interstitial phases of the system Ti_3N-Ti_3O and show that continuous solid solutions are formed in it.

However, experimental investigations of some metallide systems showed that total mutual solubility of metallides does not always presuppose the formation of continuous solid solutions between two of the constituent elements in the compounds. Hence the condition given above requires some modification. Earlier we formulated it in the following manner: Two of the constituent elements of compounds in metallide systems must be analogs (belonging to the same group) or have nearly identical metallochemical properties. This is due to the fact that such elements lie close together in the same group or the same period.

Analysis of existing experimental data confirms the above. Many known metallide systems are given in Table 42, in which the formation of continuous solid solutions, even in the absence of total mutual solubility between the paired constituent elements of these compounds, has been proved experimentally. Together with the metallide systems, the table gives data on types of chemical bonding and crystal structure, as well as the difference in atomic radius between the paired constituent elements.

In the periodic system these paired elements evidently lie close together in a given period (Si and Al, Sn and Sb, Cd and In, Ga and Zn, Pb and Bi, Cr and Mn, Co and Mn) or are elements of the same group (Sn and Ge, Cd and Zn, Cd and Hg, Ga and In, Al and Ga, P and As).

The differences in atomic radius and other metallochemical properties of such paired elements are small, because the elements lie close together in the same period or the same group of the periodic system.

Table 42 shows the following differences in atomic radius between paired elements of adjacent groups: Zn and Ga — 0%, Sn and Sb — 2%, Pb and Bi — 5%, Cd and In — 6.4%, Al and Si — 6.7%, and Cr and Mn — 2.2%. For pairs in which both elements lie in the same group, these differences are as follows: Cd and Hg — 1.9%, Al and Ga — 2.9%, Sn and Ge — 13.7%, P and As — 13.7%, and Ga and In 19.4%, which is the highest value. Thus, except for Ga and In, the difference in atomic radius between paired elements occurring close together in the periodic system is not more than 14-15%. This difference lies within the limited favoring exchange of these atoms in the structure of metallides. Hence, other conditions being equal (with identical types of chemical bonding and crystal structure) in the systems given in Table 42, continuous solid solutions are formed. This is favored by the fact that the component compounds in each system have identical types of chemical bonding and crystal structure (see Table 42).

In connection with this the question naturally arises as to why these paired elements do not form continuous mutual solid solutions in simple metallic systems. This may be explained in the following manner: Firstly, the paired elements do not all have the same crystal structure, e.g., Al and Si, Pb and Bi, Al and Ga, Zn and Ga, etc; secondly, they have different types of chemical bonding; for instance, Al has metallic bonding, Si and Ge have covalent, and Sn metallic.

All the foregoing is unfavorable to the formation of solid solutions in metallic systems, and hence many of the elements paired above form limited mutual solid solutions at best, whereas mutual solubility is nearly absent in the system Al—Si. This also emphasizes the extent to which difference in type of chemical bonding limits solubility in solid solutions or even precludes their formation.

When these elements become components of metallides, they lose their individual properties, and the differences noted above, limiting the formation of solid solutions in simple systems, vanish in metallide ones. Thus Al and Si lose their intrinsic structural characteristics and individual properties on becoming components of the aluminides and silicides given above (e.g., Mo_3Al and Mo_3Si). Both these compounds containing Al and Si atoms, whose radii are nearly identical, have metallic bonding; they are isomorphous. Under these conditions all prerequisites exist for total mutual solubility of these two metallides despite the fact that Al and Si in a simple system do not form mutual solid solutions. The arguments put forth above for aluminides and silicides are valid also for the other metallide systems given in Table 42.

TABLE 42. Continuous Solid Solutions of Metallide Systems Whose Constituent Elements Are Mutually Insoluble

System*		Type of chemical bonding		Type of crystal structure		Difference in atomic radius between elements, %
I	II	I	II	I	II	
Mo_3Si	Mo_3Al	Met.	Met.	β-W	β-W	Si and Al=6.7
Nb_3Sn	Nb_3Sb	»	»	β-W	β-W	Sn and Sb=2.0
Zr_5Si_3	Zr_5Al_3	»	»	Mn_5Si_3	Mn_5Si_3	Si and Al=6.7
Hf_5Si_3	Hf_5Al_3	»	»	Mn_5Si_3	Mn_5Si_3	Si and Al=6.7
Zr_2Si	Zr_2Al	»	»	$CuAl_2$	$CuAl_2$	Si and Al=6.7
Hf_2Si	Hf_2Al	»	»	$CuAl_2$	$CuAl_2$	Si and Al=6.7
SnTe	GeTe	Coval.	Coval.	NaCl	NaCl	Sn and Ge=13.7
CdTe	HgTe	»	»	ZnS	ZnS	Cd and Hg=1.9
CdTe	InTe	»	»	ZnS	—	Cd and In=6.4
CdSb	InSb	»	»	Rhomb.	—	Cd and In=6.4
GaSb	InSb	»	»	ZnS	ZnS	Ga and In=19.4
PtBi	PtPb	Met.	Met.	NiAs	NiAs	Bi and Pb=5.0
CrSi	MnSi	»	»	Cub. (FeSi)	Cub. (FeSi)	Cr and Mn=2.2
MnSi	CoSi	»	»	Same	Same	Mn and Co=5.0

*I refers to compound I of system; II refers to compound II of system.

In addition to the above, it should be kept in mind that when some metallides are formed, the radii of their component atoms may change. Thus, for instance, it was shown in an investigation of the character of chemical bonding in transition metal aluminides and silicides [253] that in some such compounds the sum of atomic radii is less than the sum of average atomic radii, according to Goldschmidt. There are also other examples attesting to change in atomic radii on compound formation, particularly in cases where covalent and other intermediate types of chemical bonding appear.

These peculiarities must be taken into account also when questions of metallide interactions are being considered as proof of the presence or absence of solid solutions in metallide systems.

Developing these arguments, one may expect, other conditions being equal, the formation of continuous solid solutions between metallides containing the following pairs of elements: Ag and Cd, Cu and Zn, Hg and Tl, Tl and Pb, Ga and Ge, In and Sn, U and W, and some others. On the other hand, one cannot expect total exchange in solid solutions of metallides in systems where the compounds contain elements having markedly different metallochemical properties. This thesis is valid whether these elements lie in the same group or merely close together in the same period. Thus, for instance, one cannot expect continuity in metallide solid solutions containing the following pairs of elements: Si and C (silicides−carbides), Si and Pb (silicide− plumbides), Be and B (beryllide−borides), Al and B (aluminide−borides), B and C (boride−carbides), etc. The elements given above, although they occur in the same group or are adjacent, differ so much in atomic radius, electronegativity, and ionization potential (see Table 1 and Figs. 1 and 2) that atoms of these elements cannot exchange on formation of continuous metallide solid solutions. Hence it must be presumed that in such systems one can expect the formation only of limited solid solutions or, in the extreme case, no solid solutions. This should explain the discontinuity or total absence of solid solutions in systems of borides and carbides, carbides and silicides, aluminides and borides, and borides and beryllides.

The presence of identical atomic ratios as a condition of total mutual solubility of metallides is based on the fact that such compounds, as a rule, have identical crystal structures and chemical bonding. When the atomic ratios are different and the above three conditions are not all met, these compounds give only limited solid solutions. However, there are cases where continuous solid solutions are formed even when the atomic ratios in the compounds are different. This was shown in the case of reactions between electron compounds [143] and between certain so-called heterovalent semiconductor compounds [105].

The systems [105] $CdTe-CuInTe_2$, $CdTe-AgInTe_2$, $GaAs-ZnGeAs_2$, and $InAs-ZnGeAs_2$, as well as the systems [254, 255] $PbS-AgSbS_2$, $PbS-AgBiSe_2$, $PbSe-AgSbTe_2$, $PbSe-AgSbSe_2$, etc., may be cited as examples of continuous solid solutions in heterovalent systems. The phase diagrams of the last two systems, in which continuous solid solutions are formed [254],are shown in Fig. 96.

Based on the above four theses regarding the formation of continuous solid solutions, one can understand the limitation or absence of mutual solubility in metallide systems. Saturated solid solutions of metallides appear when the compounds are nonisomorphous, polymorphic (nonisomorphous) modifications are present, or the compounds have different types of chemical bonding.

In the extreme case, where the metallochemical properties of the compounds are quite different, the conditions for mutual solubility are not met in reactions between metallides, and solid solutions of appreciable concentrations are not formed in these systems. Such cases occur in reactions between compounds with different types of chemical bonding: $CuAl_2$ and $AlSb$, Ni_3Al and NiO, etc.

These rules are important for further development of metal chemistry and the physicochemical theory of metallide interactions; they are required for predicting the character of interaction of many binary metallide systems not yet studied. To some degree, such predictions will be useful to the experimental investigator in choosing subjects of study and calculating the results to be expected.

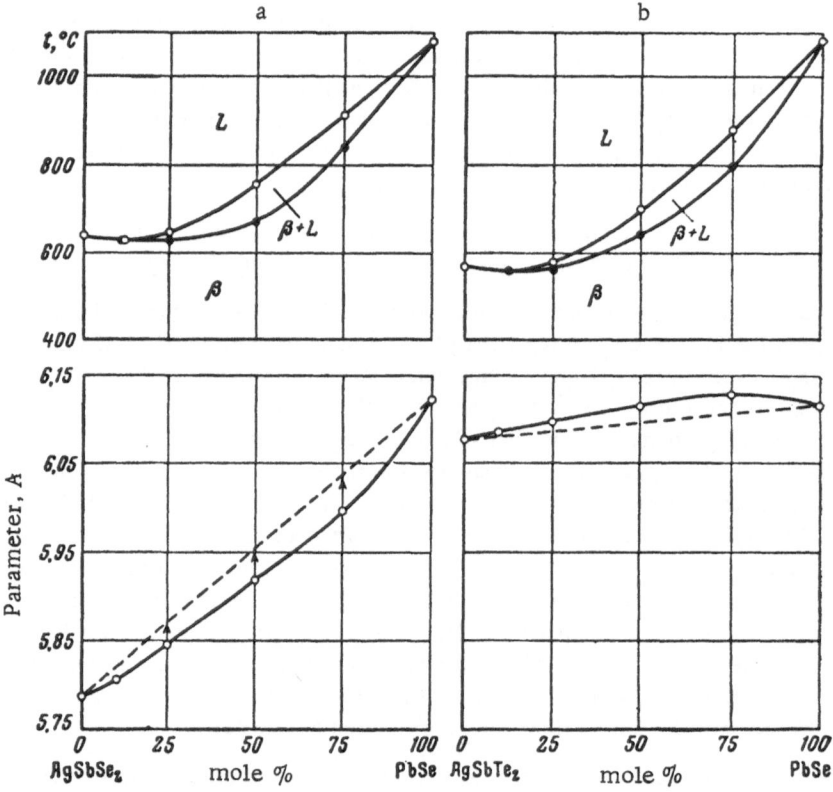

Fig. 96. Diagrams of fusibility and lattice parameter at 25° versus composition for systems: a) $AgSbSe_2-PbSe$; b) $AgSbTe_2-PbSe$.

In some sections of this book (see above) we gave a number of metallide systems in which the formation of continuous or limited solid solutions might be expected from the viewpoint of metallide interaction theory. This referred to systems composed of borides, silicides, carbides, and nitrides, as well as superconductor and semiconductor compounds. For predicting the character of their reactions, the number of possible systems may be increased considerably if it is taken into account that the total number of metallides exceeds 4000. Here there may be systems composed of metallides of alkali and alkaline-earth metals, including an interesting new class of compounds — beryllides, many compounds of aluminum and its analogs (aluminides, gallides, indides, thallides), compounds based on transition metals, and also many compounds which we classify as phosphides, arsenides, antimonides, borides, carbides, and silicides. In this respect, systems of compounds of Group VI elements — oxides, sulfides, selenides, and tellurides — in which not only ionic bonding, but also covalent and even metallic or mixed types of bonding are manifested, may constitute an independent group of metallide systems. Here it will be noted that compounds with covalent and metallic bonding, containing such elements, increase in number with the transition $O \rightarrow S \rightarrow Se \rightarrow Te$, on the one hand, and with increasing concentration of metal atoms in such compounds, on the other. A striking example in this respect is the series of titanium—oxygen compounds $TiO_2 \rightarrow Ti_2O_3 \rightarrow TiO \rightarrow Ti_3O \rightarrow Ti_6O$.

As is evident, the first two of these, TiO_2 and Ti_2O_3, are typical ionic compounds; TiO has semiconductor and, to some degree, metallic properties, whereas the newly discovered suboxides Ti_3O and Ti_6O [48] are typical metallic compounds.

In connection with the discussion of suboxide formation in metal—oxygen systems, attention should be directed toward the possible existence of such suboxides in other metallic systems. Most of these oxides are formed either by low-temperature reactions (oxidation) between metals and oxygen or by reduction of normal (higher) oxides.

Since they are formed in the solid state and decompose on heating in the solid state, they may be regarded as Kurnakov compounds. Many such compounds — suboxides — have metallic bonding; hence they may be called oxides in analogy with the names adopted above (silicides, carbides, nitrides, etc.).

Suboxides of Group V metals — vanadium, niobium, and tantalum — and Group VI ones — chromium and tungsten — whose compositions have not been precisely determined as yet, belong to this type. Such compounds are described in [43]; moreover, new papers on metal suboxides have appeared recently.

Thus, for instance, there are statements in the literature [256] regarding the possible formation of vanadium suboxides V_6O and V_3O and the existence of a homogeneous region of solid solutions between them. Together with vanadium monoxide (VO), these compounds are formed in the reduction of V_2O_5 by hydrogen at 800°.

Interesting investigations were published recently [257] on a niobium suboxide of composition Nb_6O. The authors prepared this compound by low-temperature (350-450°) oxidation of niobium. In this case it was noted that when niobium is oxidized at temperatures below 350°, this suboxide is formed in the amorphous state, becoming crystalline on further oxidation. On oxidation at 400-450° Nb_6O is formed directly in the crystalline state. According to preliminary data, it has a tetragonal structure. At higher temperatures (above 450°) Nb_6O becomes unstable and goes over to higher oxides — Nb_2O, NbO, and Nb_2O_5.

According to [257], the general scheme of reactions of niobium with oxygen at moderate temperatures is as follows:

$$Nb + O_2 \rightarrow Nb(O_{solid\ solution}) \rightarrow Nb_6O \rightarrow Nb_2O \rightarrow NbO \rightarrow Nb_2O_5.$$

The conditions of formation of tantalum suboxides were studied in the same paper. It was shown that when tantalum reacts with oxygen at 300-500°, suboxide phases are formed. One such phase corresponds to the compound Ta_6O and has a tetragonal structure. It was proved that when tantalum is oxidized at temperatures above 800° at low oxygen pressure, a new suboxide of composition Ta_2O is formed.

The order in which oxides are formed by tantalum during its oxidation may be expressed by the analogous scheme

$$\mathrm{Ta + O_2 \rightarrow Ta(O_{solid\ solution}) \rightarrow Ta_6O \rightarrow Ta_2O \rightarrow TaO \rightarrow Ta_2O_5.}$$

Data on suboxides of Group VI metals (Cr, Mo, and W) are less definite. One may say that there are indications that a chromium compound of composition Cr_3O may exist [43]. The presence of this suboxide was inferred mainly from x-ray data; it is said to have a β-W-type crystal structure [258].

There is a report [259] on the possible existence of the compound CrO, which is not yet confirmed. There is no concrete information on molybdenum suboxides in the literature. There are indications of the possible existence of tungsten suboxides in the tungsten—oxygen system. Thus, for instance, the possible existence of the compound W_3O with the β-W structure is noted in [260], whereas the author denies the existence of the modification β-W itself.

In connection with the study of possible reactions of suboxide formation in the tungsten—oxygen system, the papers [261, 262] are of interest. The thermodynamics of lower tungsten oxides is investigated in the first, whereas the second deals with the equilibrium diagram of the tungsten—oxygen system.

In [262] reactions between oxygen and tungsten in the solid state were studied and the free energies of formation of tungsten oxides, including lower oxides, determined. As a result, the authors established the existence of the tungsten compounds W_3O, WO_2, $W_{20}O_{58}$, $W_{18}O_{49}$, and WO_3 and determined their regions of existence in the phase diagram of this system. Based on the temperature conditions of their stable states, a new equilibrium diagram of the tungsten—oxygen system is given in this paper (Fig. 97). Although preliminary and in many respects schematic, it is interesting from the standpoint of new views on the formation of metal suboxides, particularly those of tungsten, and indications of these compounds in the phase diagram of the tungsten—oxygen system.

Thus on the basis of the above materials on reactions of metals in Groups IV, V, and VI of the periodic system with oxygen, a new, large class of metallides has been established which we call suboxides. They are

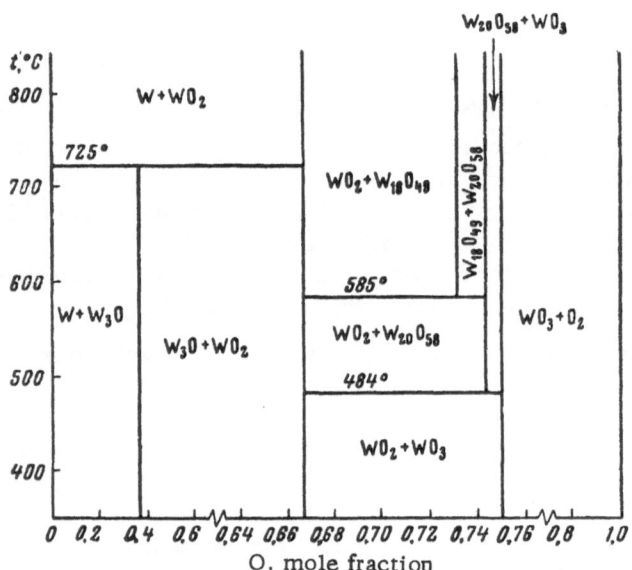

Fig. 97. Equilibrium diagram of system W—O at atmospheric pressure.

formed with various atomic ratios and correspond to the formulas Me_6O, Me_2O, MeO, etc. Their compositions do not conform to the laws of valence. Most such compounds are formed in the solid state and decompose (in the solid state) at high temperatures. They are Kurnakov compounds, and many of them have metallic bonding. One can imagine cases where solid solutions are formed between isomorphous, analogous suboxides. They are of interest for further experimental investigation.

Based on the rules considered above, one may predict many systems of metallides forming solid solutions. To illustrate where one should expect the formation of continuous or limited solid solutions in systems of various classes of compounds, one may cite certain metallide systems (Table 43). The table includes systems of semiconductor compounds of types A_IB_V and $A_{II}B_{IV}$, in which the constituents A are alkali or alkaline-earth metals, and constituents B are the elements Sb and Si. Thus they are antimonides and silicides of these electropositive metals. Besides these compounds, the table includes certain beryllides and aluminides with metallic bonding, having a high content of Be or Al atoms.

Besides the compositions of the compounds, the table gives the types of crystal structure and chemical bonding and the results that one should expect when investigating the formation of solid solutions in such systems. The compositions and some properties of the compounds are described in the papers and books cited in Table 43; however, no systems composed of them have been studied.

Based on the theses set forth, we predict the presence of continuous or limited solid solutions in these systems. Thus, for instance, in the series of alkali metal (Li, Na, K) arsenides one should expect the formation of limited solid solutions. This follows from the fact that these metals are quite different from one another in their metallochemical properties (see Table 1 and Figs. 1 and 2), especially atomic radius. Only K, Rb, and Cs have nearly identical properties, and one may expect the formation of continuous solid solutions in systems of their antimonides.

One may reach similar conclusions regarding the systems of magnesium, beryllium, and aluminum compounds listed in Table 43. Investigations of the interactions of relatively new heat-resistant and high-temperature oxidation-resistant compounds—beryllides of a number of transition metals — are very interesting from the theoretical and practical viewpoints [225]. The compositions of such little-studied compounds are given in the table. Our predictions regarding beryllide systems are rather arbitrary in some cases, since the structure types of these compounds are unknown. This complicates their further study, making it all the more difficult to determine the character of their reactions.

Table 43 could be extended by adding other little-known or entirely unknown metallide systems. One might also consider systems of intermetallides, sulfides, oxides, and hydrides. Literature data on hydrides [264, 265] show some possibilities of existence of mutual hydride solid solutions. These questions require further investigation.

In connection with the participation of metallides in the structure of complex alloys, methods of investigating such simple and many-component alloys in equilibrium with these metallides or their solid solutions are quite important. In simple systems this is done by studying so-called quasi-binary (Me—AB) and quasi-ternary (Me—AB—AC) systems. Investigation of systems containing more than four components (including compounds) is more complicated. In these cases one must resort to complex methods for studying them and representing phase equilibria geometrically. These methods for studying equilibria in many-component systems are the subject of many investigations [14, 15, 266, 267].

Owing to the complexity of methods of multidimensional geometry in the study of equilibria in many-component systems, including those with metallides, simplified methods of representation are used. One such method is that of geometrical representation of partial many-component systems containing more than four components [220]. In this case one of the compositions of the many-component solid solutions is taken arbitrarily as a component, and the equilibrium of this solid solution with the metallides formed in the complex system is studied.

TABLE 43. Metallide Systems Where Solid Solution Formation Should be Expected

Metallide system I — II	Structure type		Type of chemical bonding		Expected results of interaction in system
	I	II	I	II	
Alkali metal compounds [43, 97, 100, 101, 103, 230]					
Li_3Sb—Na_3Sb	α-hex.(Na_3As) β-cub.(BiF_3)	hex. (Na_3As)	Coval. (semip.)	Coval. (semip.)	Lim. solubility
Li_3Sb—K_3Sb	Same	Same	Same	Same	Same
Li_3Sb—Rb_3Sb		Unk.	»	»	»
Li_3Sb—Cs_3Sb	»	»	»·	»	»
Na_3Sb—K_3Sb	Hex.	Hex.	»	»	»
Na_3Sb—Rb_3Sb	»	Unk.	»	»	»
Na_3Sb—Cr_3Sb	»	»	»	»	»
K_3Sb—Rb_3Sb	»	»	»	»	Cont. soly.
K_3Sb—Cs_3Sb	»	»	»	»	»
Rb_3Sb—Cs_3Sb	»	»	»	»	»
Magnesium compounds [43, 97, 101, 103]					
Mg_2Si—Mg_2Ge*	Cub. (CaF_2)	Cub. (CaF_2)	Coval. (semip.)	Coval. (semip.)	Cont. soly.
Mg_2Si—Mg_2Sn	Same	Same	Same	»	Lim. soly.
Mg_2Si—Mg_2Pb	»	»	»	Met.	»
Mg_2Si—Ca_2Si	»	Tetrag.	»	Coval. (semip.)	»
Mg_2Ag—Mg_2Au	Hex.	Hex.	Met.	Met.	Cont. soly.
Beryllium compounds [43, 223]					
$MoBe_{12}$—$NbBe_{12}$	Unk.	Hex.	Met.	Met.	Cont. soly.
$MoBe_{12}$—$TaBe_{12}$	»	Tetrag.	»	»	Same
$MoBe_{12}$—$TiBe_{12}$	»	Unk.	»	»	»
$NbBe_{12}$—$TaBe_{12}$	Hex.	Hex.	»	»	»
$NbBe_{12}$—$TiBe_{12}$	»	Unk.	»	»	»
$TaBe_{12}$—$TiBe_{12}$	Unk.	»	»	»	»
Nb_2Be_{17}—Ta_2Be_{17}	Hex.	Hex.	»	»	»
Nb_2Be_{17}—Zr_2Be_{17}	»	»	»	»	Lim. soly.
Ta_2Be_{17}—Zr_2Be_{17}	»	»	»	»	Same
$MoBe_2$—$TiBe_2$	Hex. ($MgZn_2$)	Cub. ($MgCu_2$)	»	»	»
$MoBe_2$—$NbBe_2$	»	Unk.	»	»	—
$NbBe_2$—$TiBe_2$	»	»	»	»	—
$ThBe_{13}$—UBe_{13}**	Face-centered	Face-centered	»	»	Cont. soly.
Aluminum compounds [43, 231]					
$MoAl_{12}$—WAl_{12}	Cub. (CsCl)	Cub. (CsCl)	Met.	Met.	Cont. soly.
$NbAl_3$—$TiAl_3$	Tetrag.	Tetrag.	»	»	Same
$NbAl_3$—$ZrAl_3$	»	»	»	»	»
$TiAl_3$—$ZrAl_3$	»	»	»	»	»

*System studied recently [187].

**The beryllide system $ThBe_{13}$—UBe_{13} was studied recently [263]. As should have been expected from the main conditions of formation of solid solutions, this system corresponds to continuous solid solutions. In the cited article [263] the authors give diagrams of hardness and lattice parameter versus composition, in which the properties vary continuously with composition; this attests to total mutual solubility of these two metallides.

Thus, for instance, a quaternary Ni—Cr—W—Mo solid solution, whose composition was such that it had a homogeneous structure at all temperatures and did not undergo any solid-state transformations, was taken as the starting point for investigating and representing the nickel-based partial seven-component system Ni—Cr—W—Mo—Nb—Ti—Al. The remaining three elements (Nb, Ti, Al) were added to this system in the form of the metallides $NbNi_3$, $TiNi_3$, and Ni_3Al. Then the compositions of alloys of the nickel-based seven-component system could be represented in equilibrium in the form of a tetrahedron, as shown schematically in Fig. 98. In this figure, the composition of the four-component solid solution is located at one vertex of the tetrahedron; along the edges are plotted the quasi-binary systems: four-component solid solution—metallides $TiNi_3$, $NbNi_3$, and Ni_3Al and metallide—metallides $TiNi_3$—$NbNi_3$, $NbNi_3$—Ni_3Al, and Ni_3Al—$TiNi_3$; the corresponding quasi-ternary systems are plotted on the faces, whereas the compositions of alloys of the seven-component solid solution and phases based on the three metallides $TiNi_3$, $NbNi_3$, and Ni_3Al, chosen for investigation, are given in the body of the tetrahedron.

In the three-dimensional diagram this method of representation brings out: a broad region of seven-component nickel γ-solid solution, a region of β-continuous solid solutions based on the system $TiNi_3$—$NbNi_3$, regions of limited solid solutions of α- and η-systems based on $TiNi_3$, Ni_3Al, and $NbNi_3$, and also the corresponding two- and three-phase regions bounding single-phase regions of alloys of this system.

This method of representing equilibria in many-component metal systems containing metallides and solid solutions based on them is used to study partial equilibrium diagrams of many-component systems and to represent the compositions of complex alloys. The method may be used to develop new complex alloys with optimum physical and other properties. This is shown by the development of new high-strength, heat-resistant alloys based on nickel, titanium, and other metals, as set forth in [220].

Such methods of representing the results of reactions in many-component systems based on heat-resistant alloys are suitable also for other complex systems containing many metallides. A striking example of the application of this method to a many-component metallide system consisting of five semiconductor compounds is set forth in [254]. The author, who did not intend to study the entire five-component system $PbTe$—$PbSe$—$AgSbSe_2$—$AgSbTe_2$—$AgBiSe_2$, represented it in the form of a three-dimensional section of a hexatope, drawn through two

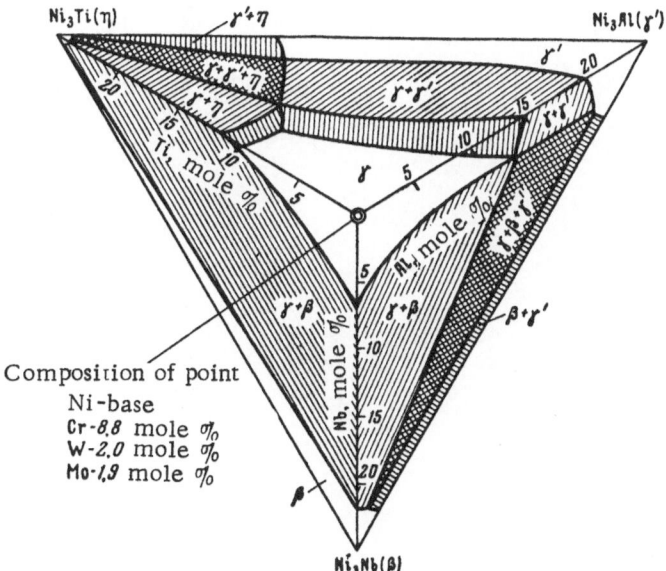

Fig. 98. Schematic representation of equilibrium of partial, Ni-based, seven-component system Ni—Cr—W—Mo—Nb—Ti—Al at 1200°.

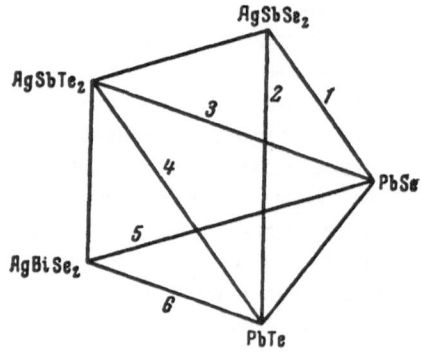

Fig. 99. Six quasi-binary systems in the five-component system PbTe−PbSe−$AgSbSe_2$−$AgSbTe_2$−$AgBiSe_2$.

binary and three ternary compounds (Fig. 99). In this complex system the author studied six secondary quasi-binary systems, two of which are shown in Fig. 96.

Numerous other many-component metallide systems containing a large number of compounds — borides, carbides, silicides, nitrides, etc. — may be represented analogously. As yet, no systematic work along this line has been done.

The importance of developing these investigations in the field of metal chemistry will be noted. Study of the character of metallide reactions in simple and many-component systems is certainly of considerable interest scientifically and also practically for developing new complex inorganic materials with special physical properties.

CONCLUSION

The investigation of inorganic compounds of metals with metals and nonmetals is interesting from the viewpoint of the theory of formation of chemical compounds and the practical application of these materials.

The study of such compounds, which we call metallides, and the character of reactions between them is one of the important problems of metal chemistry.

We have considered questions of the formation of chemical compounds of metals and their classification, taking the nature of their chemical bonding into account.

The main factors brought out earlier, determining the character of metallide interactions, were further investigated theoretically and experimentally with regard to equilibrium in simple and many-component metallide systems.

The rules of formation of continuous and limited metallide solid solutions, established in this book, enable one to predict the character of reactions between metallides in many systems not yet studied and to solve problems of practical application of new metallide-based materials.

The author thanks Corresponding Members, Acad. Sci. USSR, N. V. Ageev and N. E. Alekseevskii, Doctor of Chemical Sciences N. Kh. Abrikosov, Doctor of Technical Sciences B. K. Vul'f, and Candidates of Technical Sciences P. B. Budberg, E. I. Gladyshevskii, P. I. Kripyakevich, E. E. Cherkashin, N. M. Matveeva, and Ya.A. Ugai for examining the manuscript and for comments and additions made by them, and also O. V. Ozhimkova for her assistance in preparing the monograph for publication.

LITERATURE CITED

1. M. V. Lomonosov. Complete Works, Vol. I: Works on Physics and Chemistry, Acad. Sci. USSR Press(1950).
2. M. V. Lomonosov. Complete Works, Vol. V: Works on Mineralogy, Metallurgy, and Mining, Acad. Sci. USSR Press (1950).
3. A. I. Sherer. Guide for Teaching Chemistry, St. Petersburg, Medical Printing House (1803).
4. B. N. Menshutkin. Izv. Inst. Fiz.-Khim. Analiza Akad. Nauk SSSR 8:373 (1936).
5. C. J. Karsten. Handbuch der Eisenhüttenkunde, Vol. 1, Berlin (1841).
6. A. Matthiessen. "Report on the chemical nature of alloys," Reports of 33rd Meeting of the Brit. Assn. Adv. Science (1863), p.37.
7. D. I. Mendeleev. Fundamentals of Chemistry, Vols. I and II, Publ. No. 9, GIZ (1928).
8. N. S. Kurnakov. Selected Works, Vol. II, Khimizdat (1939).
9. W. Gibbs. Thermodynamic Works [Russian translation edited by V. K. Semenchenko], Gostekhteoretizdat (1950).
10. G. A. Tammann. Lehrbuch der Metallographie, Leipzig, Leopold Voss (1914).
11. W. C. R. Austen. An Introduction to the Study of Metallurgy, Second edition, London (1910).
12. H. LeChatelier. Compt. Rend. 130:85 (1900).
13. M. Giua. Chemical Combination Among Metals, Philadelphia, P. Blackiston's Son and Co. (1918).
14. N. S. Kurnakov. Introduction to Physicochemical Analysis, Acad. Sci. USSR Press (1940).
15. V. Ya. Anosov and S. A. Pogodin, Basic Principles of Physicochemical Analysis, Acad. Sci. USSR Press (1947).
16. A. A. Bochvar. Metallography, Metallurgizdat (1956).
17. Ya. S. Umanskii, V. N. Finkel'shtein, M. N. Blanter, et al. Physical Fundamentals of Metallography, Metallurgizdat (1955).
18. R. Voge. Die heterogenen Gleichgewichte, Leipzig, Akademie Verlag (1937).
19. F. N. Rhines. Phase Diagrams in Metallurgy. Their Development and Applications, New York, McGraw-Hill Book Company (1956) [Russian translation, Metallurgizdat (1960)].
20. W. Hume-Rothery and G. V. Raynor. Structure of Metals and Alloys [Russian translation], Metallurgizdat (1959).
21. C. S. Barrett. Structure of Metals, New York, McGraw-Hill Book Company (1943).
22. F. Seitz. The Physics of Metals, New York, McGraw-Hill Book Company (1943).
23. L. S. Darken and R. W. Gurry. Physical Chemistry of Metals, New York, McGraw-Hill Book Company (1953). [Russian translation, Metallurgizdat (1960).
24. N. V. Ageev. Chemistry of Metallic Alloys, Acad. Sci. USSR Press (1941).
25. N. V. Ageev. Nature of Chemical Bonding in Metallic Alloys, Acad. Sci. USSR Press (1947).
26. I. I. Kornilov. Usp. Khim. 21(6):1045 (1952).
27. I. I. Kornilov. Izv. Akad. Nauk SSSR, Otd. Khim. Nauk, No. 5:795 (1953).
28. I. I. Kornilov. Izv. Akad. Nauk SSSR, Otd. Khim. Nauk, No. 4:397 (1957).
29. National Phys. Lab. Symposium, No. 9. The Physical Chemistry of Metallic Solutions, and Intermetallic Compounds, Vol. 1, London, (1959).
30. L. Pauling. Nature of the Chemical Bond [Russian translation], Goskhimizdat (1947).
31. W. Gordy and W. Thomas. J. Chem. Phys. 24:439 (1956).
32. M. Haissinsky. J. Phys. Radium 7(1):7 (1946).
33. S. S. Batsanov. Electronegativity of Elements and the Chemical Bond, Acad. Sci. USSR Press, Siberian Branch (1962).
34. Ya. K. Syrkin. Usp. Khim. 31, (4): 397 (1962).
35. V. N. Spiridonov and V. M. Tatevskii. Zh. Fiz. Khim. 37(5): 994 (1963).

36. Commission for Nomenclature of Chemical Compounds, Acad. Sci. USSR, Project on Nomenclature of Inorganic Compounds, Acad. Sci. USSR Press (1959).

37. N. S. Kurnakov. Zh. St. Petersburg. Politekhn. Inst. (1914); Zh. Russk. Fiz.-Khim. Obshch. 47:871 (1915).

38. F. Weibke, Z. Elektrochem. 44:209, 263 (1938).

39. E. Zintl. Z. Angew. Chem. 52, 1 (1939).

40. I. I. Kornilov and B. K. Vul'f. Usp. Khim. 28 (9):1086 (1959).

41. G. B. Bokii, B. K. Vul'f, and N. L. Smirnova. Zh. Strukt. Khim. 2 (1):74 (1961).

42. G. V. Samsonov. Powder Metallurgy, No. 2, Acad. Sci. Ukr.SSR Press (1959).

43. M. Hansen and K. Anderko. Structure of Binary Alloys [Russian translation], Metallurgizdat (1962).

44. F. N. Rhines and J. B. Newnik. Trans. Am. Soc. Metals 45:1029 (1953).

45. I. I. Kornilov and N. M. Matveeva. Dokl. Akad. Nauk SSSR 139 (4):880 (1961).

46. I. I. Kornilov, E. N. Pylaeva, and M. A. Volkova. Dokl. Akad. Nauk SSSR 137 (3):599 (1961).

47. I. I. Kornilov, E. N. Pylaeva, and M. A. Volkova. Collection: Titanium and Its Alloys, Vol. X, Acad. Sci. USSR Press (1963), p. 74.

48. I. I. Kornilov and V. V. Glazova. Dokl. Akad. Nauk SSSR 150 (2):313 (1963).

49. S. T. Konobeevskii. Izv. Sektora Fiz.-Khim. Analiza 16 (1):19 (1943).

50. O. Kubaschewski and H. Evans. Thermochemistry in Metallurgy [Russian translation], IL (1954).

51. Yu. M. Golutvin. Heats of Formation and Types of Chemical Bonding in Inorganic Crystals, Acad. Sci. USSR Press (1962).

52. N. V. Ageev. Zh. Neorgan. Khim. 3 (3) (1958).

53. G. V. Samsonov, L. Ya. Markovskii, A. F. Zhigach, and M. G. Valyashko. Boron, Its Compounds and Alloys, Acad. Sci. Ukr.SSR Press (1960).

54. G. V. Samsonov and Ya. S. Umanskii. Hard Compounds of Refractory Metals, Metallurgizdat (1957).

55. G. V. Samsonov. Silicides and Their Engineering Applications, Acad. Sci. Ukr.SSR Press (1959).

56. G. V. Samsonov and L. L. Vereikina. Phosphides, Acad. Sci. Ukr.SSR Press (1961).

57. G. V. Samsonov and S. V. Rodzinovskaya. Usp. Khim. 30 (1):6 (1961).

58. B. F. Ormont. Structures of Inorganic Compounds, GONTI (1950).

59. G. B. Bokii. Introduction to Crystal Chemistry, Moscow State University Press (1954).

60. P. I. Kripyakevich and E. E. Cherkashin. "Classification of binary intermetallic phases," Izv. Sektora Fiz.-Khim. Analiza Akad. Nauk SSSR 24:59 (1954).

61. E. S. Makarov. Zh. Neorgan. Khim. 1:1583 (1956).

62. V. P. Zhuze (ed.). Collection: Semiconductor Materials. Questions of Chemical Bonding [Russian translation], (1960).

63. G. S. Zhdanov. Solid State Physics, Moscow State University Press (1962).

64. P. B. Beck. Electronic Structure and Chemistry of Transition Metal Alloys, New York, Interscience Publishers, Inc. (1963).

65. E. Mooser and W. Pearson. Collection: Semiconductor Materials [Russian translation], IL (1960), p. 183.

66. F. Laves. Metallwirtschaft 15:631 (1936).

67. U. Dehlinger and G. Schulze. Z. Metallk. 33 (4):157 (1941).

68. G. Schulze. Z. Elektr. und Angew. Phys. Chem. 45(12):849 (1939).

69. P. I. Kripyakevich and E. E. Cherkashin. Usp. Khim. 19 (3):361 (1950).

70. G. B. Bokii and É. E. Vainshtein. Dokl. Akad. Nauk SSSR 40 (6):262 (1943).

71. R. L. Berry and G. V. Raynor. Acta Cryst. 6 (2): 178 (1953).

72. F. Laves. "Theory of Alloy Phases," Am. Soc. Metals (1956), pp. 124-198.

73. G. Schulze. Z. Krist. 111 (4): 249 (1959).

74. E. S. Makarov. Structure of Solid Phases with a Variable Number of Atoms in the Unit Cell, Acad. Sci. USSR Press (1947).

75. P. Esslinger and K. Schubert. Z. Metallk. 48 (3):126 (1957).

76. G. V. Samsonov. Dokl. Akad. Nauk 93, (4):689 (1953).

77. G. V. Samsonov and B. M. Tsarev. High-Temperature Metal — Ceramic Materials, Acad. Sci. Ukr.SSR Press (1961).

78.   R. Kieffer and P. Schwarzkopf. Hard Alloys [Russian translation], Metallurgizdat (1957).

79.   V. I. Prosvirin. Vestn. Metalloprom. No. 12 (1937).

80.   K. P. Romadin. Electrolytic Transport in Metallic Liquid and Solid Solutions, Trans. N. E. Zhukov Air Force Engineering Academy, No. 167 (1947).

81.   J. Bernal. Collection: Metal Physics [Russian translation], GTTI (1933), p. 141.

82.   L. N. Guseva and B. I. Ovechkin. Dokl. Akad. Nauk SSSR 112 (4):681 (1957).

83.   K. N. Davydov and P. V. Gel'd. Fiz. Metal. i Metalloved. 2 (1):192 (1956).

84.   V. L. Ginzburg. Superconductivity, Acad. Sci. USSR Press (1946).

85.   B. T. Matthias. Progress in the Low-Temperature Physics, Vol. 11, Interscience Publishers, Inc. (1955).

86.   N. E. Alekseevskii. Zh. Éksperim. i Teor. Fiz. 19 (4):358 (1949); 22 (3):200 (1952).

87.   N. E. Alekseevskii, G. S. Zhdanov, and N. N. Zhuravlev. Zh. Éksperim. i Teor. Fiz. 25, (1):123 (1953).

88.   N. N. Zhuravlev and G. S. Zhdanov. Zh. Éksperim. i Teor. Fiz. 28 (2):228 (1955).

89.   M. Tannenbaum and W. Wright. J. Metals 14 (5):367 (1962).

90.   B. T. Matthias, T. H. Geballe, and V. B. Compton. Rev. Mod. Phys. 35 (2): 414 (1963).

91.   B. M. Roberts. "Superconductive Mateerials and Some of Their Properties," General Electric Company Research Lab., Rept. No. 63 (March 1963).

92.   J. K. Hulm and R. D. Braugher. Phys. Rev. 123 (5):1569 (1961).

93.   B. T. Matthias, V. D. Compton and Corenzwit. J. Phys. Chem. Solids 19(1/2) (1961).

94.   T. B. Reed, H. G. Gatos, W. J. Lafleur, and J. T. Roddy. Advanced Electric Mat. Conf. AIME (August 1962).

95.   M. D. Banus, T. B. Reed, H. G. Gatos, M. C. Lavine, and J. A. Kafalas, J. Phys. Chem. Solids, 23, 971, July 1962.

96.   N. E. Alekseevskii, E. M. Savitskii, V. V. Baron, and Yu. V. Efimov. Dokl. Akad. Nauk SSSR 145 (1):82 (1962).

97.   A. F. Ioffe, Semiconductors in Modern Physics, Acad. Sci. USSR Press (1954).

98.   A. F. Ioffe: Collection: Semiconductors in Science and Engineering, Acad. Sci. USSR Press, Vol. I (1957), Vol. II (1958).

99.   C. Goodman. New Semiconductor Materials [Russian translation], IL (1958).

100. R. A. Smith (ed.). Semiconductors, New York, Cambridge University Press [Russian translation, IL (1962)].

101. Izv. Akad. Nauk SSSR, Ser. Fiz. 21 (1):141 (1957).

102. Collection: Questions of Semiconductor Metallurgy and Physics, Acad. Sci. USSR Press (1959).

103. N. B. Hannay, (ed.). Collection: Semiconductors, New York, Reinhold Publishing Corp. [Russian translation, IL (1962)].

104. N. N. Sirota (ed.). Collection: Physics and Physicochemical Analysis, Moscow, NTO Nonferrous Metallurgy Press (1957), pp. 117-134.

105. N. A. Goryunova. Chemistry of Diamond-Like Semiconductors, Leningrad University Press (1963).

106. N. S. Kurnakov and N. I. Stepanov. Zh. Russk. Fiz.-Khim. Obshch. 37 (3):568 (1905).

107. N. I. Stepanov. Electrical Conductivity of Metallic Alloys in Connection with the Electron Theory, St. Petersburg (1911).

108. C. H. Goodman. J. Phys. Chem. Solids 6(4):305 (1958).

109. W. D. Lawson and E. Nielsen. Preparation of Single Crystals, London, Butterworths Scientific Publications (1958).

110. L. D. Dudkin and A. P. Ostranitsa. Dokl. Akad. Nauk SSSR 124 (1):34 (1959).

111. I. I. Kornilov. Dokl. Akad. Nauk SSSR 81 (4):597 (1951).

112. I. I Kornilov. Dokl. Akad. Nauk SSSR 106 (3):476 (1956).

113. I. I. Kornilov. Solid Solutions in Metallic Compounds, National Phys. Lab. Symposium, 9, London, (1959).

114. B. K. Vul'f. Ternary Metallic Phases in Alloys, Moscow, N. E. Zhukov. Air Force Engineering Academy Press (1959).

115. N. A. Goryunova, and N. N. Fedorova. Dokl. Akad. Nauk SSSR 90 (6):1099 (1953).

116. N. A. Goryunova. Substitutional Solid Solutions in Compounds with the Zinc Blende Structure, Transactions of the Conference on Semiconductor Materials, 1954, Acad. Sci. USSR Press (1955).

117.   N. A. Goryunova. Izv. Akad. Nauk SSSR, Ser. Fiz. 21 (1):120 (1957).

118.   R. B. Kotel'nikov. Zh. Neorgan. Khim. 3 (4):841 (1958).

119.   B. Post, F. Glaser, and D. Moskowitz. Acta Met. 2 (1):20 (1954).

120.   G. A. Meierson, G. V. Samsonov, R. B. Kotel'nikov, et al. Zh. Neorgan. Khim. 3 (4):898 (1958).

121.   I. I. Kornilov and E. N. Pylaeva. Dokl. Akad. Nauk SSSR 97 (3):455 (1954).

122.   I. I. Kornilov and N. M. Matveeva. Dokl. Akad. Nauk SSSR 98 (5):787 (1954).

123.   E. I. Gladyshevskii and E. E. Cherkashin. Zh. Neorgan. Khim. 1 (6):1394 (1956).

124.   E. I. Gladyshevskii. Author's abstract of dissertation, L'vov State University (1958).

125.   V. I. Mikheeva and G. G. Babayan. Dokl. Akad. Nauk SSSR 108 (6):1086 (1956).

126.   R. Vogel and H. Klose. Z. Metallk. 45 (12):670 (1954).

127.   E. E. Cherkashin, E. I. Gladyshevskii, P. I. Kripyakevich, et al. Zh. Neorgan. Khim. 3:650 (1958).

128.   I. E. Gorshkov and N. A. Goryunova. Zh. Neorgan. Khim. 3:668 (1958).

129.   J. C. Wooley and B. A. Smith. Proc. Phys. Soc. 72:214 (1958).

130.   H. Nowotny, R. Kieffer, F. Benesovsky, C. Brukl, and E. Rudy. Monatsh. Chem. 90 (5):669 (1959).

131.   P. Duwerz and F. Odell. J. Electrochem. Soc. 97 (10):299 (1950).

132.   J. H. Westbrook. Trans. Met. Soc. AIME 215 (5):807 (October 1959).

133.   V. I. Mikheeva. Izv. Sektora Fiz.-Khim. Analiza Akad. Nauk SSSR 17 (1):174 (1949).

134.   J. Pratt and G. Raynor. Proc. Roy. Soc. A205 (1080):103(1951).

135.   J. Pratt and G. Raynor. J. Inst. Metals 79 (10):211 (1951).

136.   B. K. Vul'f. Fiz. Metal. i Metalloved. 3 (1):97 (1956).

137.   B. K. Vul'f. Usp. Khim. 29 (6):774 (1960).

138.   I. I. Kornilov. Iron Alloys, Vol. III: Alloys of the System Fe−Ni−Cr, Acad. Sci. USSR Press (1956).

139.   W: Koster and W. Gnohling. Z. Metall. 51 (7):385 (1960).

140.   I. I. Kornilov and T. T. Nartova. Dokl. Akad. Nauk SSSR 131 (4):837 (1960).

141.   I. I. Kornilov and T. T. Nartova. Dokl. Akad. Nauk SSSR 140 (4):829 (1961).

142.   C. Austin and G. Murphy. J. Inst. Metals 29 (2):327 (1923).

143.   E. I. Gladyshevskii. Izv. L'vov. Gos. Univ., No.8:111 (1957).

144.   N. V. Ageev and E. S. Makarov, Izv. Akad. Nauk SSSR, Otd. Khim. Nauk, No. 3:161 (1943).

145.   E. I. Gladyshevskii and E. E. Cherkashin. Nauchn. Zap. L'vovsk. Gos. Univ. 34 (4):51 (1955).

146.   Strukturbericht 3:312 (1937).

147.   J. H. Wernick, S. S. Hasko, and D. Dorsi. J. Phys. Chem. Solids 23:567 (June 1962).

148.   I. I. Kornilov and N. G. Boriskina. Zh. Neorgan. Khim. 9 (3):702 (1964).

149.   B. W. Lewinger. J. Metals, 5:21 192 (1953).

150.   R. J. Van Thyne, H. D. Kessler, and M. Hansen. Trans. Met. Soc. AIME 197:1209 (1953).

151.   V. N. Svechnikov, Yu. A. Kocherzhinskii, V. I. Latysheva, and V. M. Pan. Questions of Metal Physics and Metallography, No. 16, Acad. Sci. Ukr.SSR Press (1962), p. 128.

152.   I. I. Kornilov, P. B. Budberg, K. I. Shakhova, and N. A. Nedumov. Dokl. Akad. Nauk SSSR 149 (6):1340 (1963).

153.   V. N. Svechnikov and A. Ts. Spektor. Questions of Metal Physics and Metallography, No. 16, Acad. Sci. Ukr.SSR Press (1962), p. 145.

154.   I. I. Kornilov and E. N. Pylaeva. Dokl. Akad. Nauk SSSR 97 (3):455 (1954).

155.   I. I. Kornilov and E. N. Pylaeva. Zh. Neorgan. Khim. 3 (3):673 (1958).

156.   P. T. Kolomytsev. Dokl. Akad. Nauk SSSR 124 (6):1247 (1959).

157.   I. I. Kornilov and P. T. Kolomytsev. Dokl. Akad. Nauk SSSR 125 (2):325 (1959).

158.   G. Hagg and R. Kiessling. J. Inst. Metals 81 (1):57 (1952).

159.   R. Steinitz. Powder Met. Bull. 6 (1):54 (1951).

160.   J. Lafferty. Phys. Rev. 79 (1):1012 (1950).

161.   J. McMullin and J. Norton. J. Metals (II Section) 5 (9):1205 (1953).

162.   F. N. Tavadze. Chrome−Manganese Cast Irons, Author's abstract of doctoral dissertation, Georgian Polytech. Inst., Tbilisi (1947).

163.   H. Nowotny, R. Kieffer, F. Benesowsky, C. Brukl, and E. Rudy. Monatsh. Chem. 90 (2):86 (1959).

164.   E. I. Gladyshevskii, P. I. Kripyakevich, and Yu. B. Kuz'ma. Fiz. Metal. i Metalloved. 2 (3):454 (1956).

165.   R. Kieffer, F. Benesowsky, and H. Schwartz. Z. Metallk. 44(9):437 (1953).

166.   H. Nowotny, A. Nachenschalk, R. Kieffer, F. Benesowsky. Monatsh. Chem. 85 (1):241 (1954).

167.   E. Parthe and H. Nowotny. Monatsh. Chem. 86 (3):385 (1955).

168.   H. Kudielka and H. Nowotny. Monatsh. Chem. 87(3):471 (1956).

169.   An Hsi-yung. Investigation of the Phase Equilibrium Diagram of Mo−Si−Cr Alloys, Author's abstract of dissertation, Moscow Steel Inst. (1962).

170.   E. I. Gladyshevskii. Poroshkovaya Met. No. 4:46 (1962).

171.   T. A. Badaeva and L. I. Rybakova. Collection: Structure and Properties of Uranium, Thorium, and Zirconium Alloys, O. S. Ivanov (ed.), Moscow, Gosatomizdat (1963). p. 299.

172.   L. Stone and H. Margolin. J. Metals 5 (11) (1953).

173.   G. V. Samsonov, T. S. Verkhoglyadova, S. N. L'vov, and V. F. Nemchenko. Dokl. Akad. Nauk SSSR 142 (4): 862 (1962).

174.   B. T. Matthias, E. A. Wood, E. Corenzwit, and V. B. Baler. J. Phys. Chem. Solids 1:188 (January 1956).

175.   R. M. Bozorth, B. T. Matthias, and D. D. Davis. Proceedings of the Seventh International Conference on Low-Temperature Physics, University of Toronto (Canada) Press (1961), p. 385.

176.   N. E. Alekseevskii, E. M. Savitskii, V. V. Baron, and Yu. V. Efimov. Dokl. Akad. Nauk SSSR 145(1):82 (1962).

177.   T. B. Reed, H. G. Gatos, W. J. LaFleur, and J. T. Roddy. Advanced Electric Material Conf. AIME (August 1962).

178.   N. N. Zhuravlev, G. S. Zhdanov, and E. M. Smirnova. Fiz. Metal. i Metalloved. 13 (1):62 (1962).

179.   E. I. Elagina and N. Kh. Abrikosov. Dokl. Akad. Nauk SSSR 111(2):353 (1956).

180.   N. Kh. Abrikosov, A. M. Vasserman, and L. V. Poretskaya. Dokl. Akad. Nauk SSSR 123(2):273 (1958).

181.   M. L. Beglaryani and N. Kh. Abrikosov. Dokl. Akad. Nauk SSSR 128(2):345 (1959).

182.   O. G. Folberth. Z. Naturforsch., 10a:502 (1955).

183.   P. Baruch and M. Desse. Compt. Rend. Acad. Sci. 241:16 (1940).

184.   E. A. Peretti. Trans. Met. Soc. AIME 1:79 (1959).

185.   A. J. Strauss and Farnell Lynne. Quart. Progress Rept. Solid State Research (October 28, 1959).

186.   A. J. Strauss and Farnell Lynne. Quart. Progress Rept. Solid State Research (April 24, 1960).

187.   R. J. Labotz, D. Mason and D. F. O'Kane. J. Electrochem. Soc. 110(2):127 (1963).

188.   H. Nowotny. Usp. Khim. 27(8):996 (1958).

189.   G. V. Samsonov. Zh. Strukt. Khim. 4(3):395 (1963).

190.   J. R. Van Wazer. Phosphorus and Its Compounds, 2 Vols., New York, Interscience Publishers, Inc. (1958, 1961); Russian translation: Izd. IL, 1962.

191.   S. Geller and J. H. Wernick. Acta Cryst. 12(1):46 (1959).

192.   J. H. Wernick, S. Geller, and K. E. Benson. J. Phys. Chem. Solids 2/3:240 (1958).

193.   A. T. Grigor'ev and V. V. Kuprina. Zh. Neorgan. Khim. 6(8):1891 (1961).

194.   V. M. Pan. Dokl. Akad. Nauk Ukr.SSR 3:332 (1961).

195.   K. Lieser and H. Witte. Z. Metallk. 43(11):396 (1952).

196.   G. Petrow, S. Steeb, and I. Ellinghaus. J. Nucl. Mater. 4(3):316 (1962).

197.   G. Petrow and I. Kvernes. Z. Metallk. 52 (10):603 (1961).

198.   G. Petrow and R. Tank. Z. Metallk. 54(2):91 (1963).

199.   R. S. Mints, G. F. Belyaeva, and Yu. S. Malkov. Dokl. Akad. Nauk SSSR 143(4):871 (1962).

200.   M. P. Arbuzov and V. G. Chuprina. Collection: Investigations of Heat-Resistant Alloys, Vol. 8, Acad. Sci. USSR Press (1962).

201.   L. B. Dubrovskaya and P. V. Gel'd. Zh. Neorgan. Khim. 7 (1):145 (1962).

202.   H. Holleck, H. Nowotny, and F. Benesowsky. Monatsh. Chem. 94:359 (1963).

203.   F. Galasso, B. Bayles, and S. Soehle. Nature 198(8):984 (June 1963).

204.   N. Kh. Abrikosov. Dokl. Akad. Nauk SSSR 129(1):135 (1959).

205.   M. S. Mirgalovskaya and E. V. Skudnova. Zh. Neorgan. Khim. 4(5):1111 (1959).

206.   E. I. Elagina and N. Kh. Abrikosov. Zh. Neorgan. Khim. 4 (7):1638 (1959).

207.   E. I. Elagina. Collection: Transactions of the Fourth Conference on Semiconductor Metallurgy and Physics, Acad. Sci. USSR Press (1961).

208.   M. I. Zamotorin and L. N. Solov'eva. Metallurg, No. 7/2:115, 11 (1939).

209.   I. I. Kornilov and E. N. Pylaeva. Izv. Akad. Nauk SSSR, Otd. Tekhn. Nauk, No. 2:197 (1961).

210.   H. Nowotny and R. Kieffer. Metallforschung 2:19, 257 (1947).

211.   G. D. Lody and I. I. Tank. Proceedings of the Seventh International Conference on Low-Temperature Physics, University of Toronto (Canada) Press (1961).

212.   E. Rudy, H. Nowotny, F. Benesowsky, Rokieffer, and A. Neckel. Monatsh. Chem. 91(1):176 (1960).

213.   O. S. Ivanov and Z. M. Alekseeva. Collection: Structure of Alloys of Some Systems Containing Uranium and Thorium, Gosatomizdat (1961).

214.   A. Taylor. J. Metals (II Section) 8 (10):1356 (1956).

215.   H. Nowotny, R. Kieffer, F. Benesowsky, C. Brukl, and E. Rudy. Monatsh. Chem. 90 (5):669 (1959).

216.   A. Taylor. J. Metals (II Section) 9 (1):72 (1957).

217.   E. M. Savitskii. Effect of Temperature on the Mechanical Properties of Metals and Alloys, Acad. Sci. USSR Press (1957).

218.   J. H. Westbrook (ed.). Mechanical Properties of Intermetallic Compounds, New York, John Wiley & Sons, Inc. (1960). [Russian translation: Metallurgizdat, (1962)].

219.   G. V. Samsonov. Refractory Compounds, Metallurgizdat (1963).

220.   I. I. Kornilov. Physicochemical Bases of Heat Resistance in Alloys, Acad. Sci. USSR Press (1961).

221.   R. S. Mints. Collection: Investigations of Heat Resistance in Alloys, Vol. V, Acad. Sci. USSR Press (1959), p. 179.

222.   W. J. Buchler and R. C. Wiley. Trans. Am. Soc. Metals (June 1962).

223.   B. C. Giessen, H. Ibach, and N. J. Grant. Trans. Am. Natl. Soc. AIME (1963).

224.   G. Tamman and K. Dahl. Z. Anorg. Chem. 126:104 (1923).

225.   A. J. Stonehouse, R. M. Paine, and W. W. Beaver. In: Mechanical Properties of Intermetallic Compounds, New York, John Wiley & Sons, Inc. (1960), [Russian translation, Metallurgizdat (1962)], p. 222.

226.   G. V. Samsonov. Some Properties of Aluminides, Acad. Sci. Ukr. SSR Press (1961).

227.   A. S. Zaimovskii and L. A. Chudnovskaya. Magnetic Metals, Princeton, New Jersey, D. Van Nostrand Co., Inc. (1951); Gosenergoizdat (1957).

228.   R. M. Bozorth. Ferromagnetism, Princeton, New Jersey, D. Van Nostrand Co., Inc. (1951). [Russian translation, IL (1956)].

229.   Collection: Magnetic Properties of Metals and Alloys [Russian translation edited by S. V. Vonsovskii, IL (1961)].

230.   Collection: Magnetic Structure of Ferromagnetics [Russian translation edited by S. V. Vonsovskii, IL (1959)].

231.   D. P. Oxley, R. S. Tebble, and K. C. Williams, J. Appl. Phys. 34 (4) Pt. 2:1362 (1963).

232.   E. M. Savitskii, V. F. Terekhova, I. V. Burov, I. A. Markova, and O. P. Naumkin. Alloys of Rare-Earth Metals, Acad. Sci. USSR Press (1962).

233.   E. A. Skrabek and W. E. Wallau. J. Appl. Phys. 34 (4) Pt 2:1356 (1963).

234.   J. H. Wernick and S. Geller. Acta Cryst. 12:662 (September 1959).

235.   E. A. Nesbitt, H. J. Williams, J. H. Wernick, and R. C. Sherwood. J. Appl. Phys. 32:342 (1961).

236.   E. A. Nesbitt, H. J. Williams, J. H. Wernick, and R. C. Sherwood. J. Appl. Phys. 33:1674 May 1962).

237.   V. I. Arkharov. Oxidation of Metals, Metallurgizdat (1945).

238.   K. Hauffe. Reactions in Solids and on Their Surfaces, Pt. II [Russian translation, IL (1963)].

239.   I. I. Kornilov and N. M. Matveeva. Izv. Akad. Nauk SSSR, Ser. Metallurgiya i Gornoe Delo, No. 1:143 (1964).

240.   C. Hilsum and A. Rose-Innes. $A_{III}B_V$-Type Semiconductors [Russian translation], IL (1963).

241.   M. I. Aliev. Thermal Conductivity of Semiconductors, Baku, Acad. Sci. Azer. SSR Press (1963).

242.   A. V. Ioffe and A. F. Ioffe. Fiz. Tverd. Tela 2:781 (1960).

243.   Collection: Theory of Superconductivity [Russian translation edited by N. I. Bogolyubov, IL (1960)].

244. J. E. Kunzler and M. Tannenbaum. Sci. Am. 206: 206 (June 1962).

245. B. T. Matthias, T. H. Geballe, and V. B. Compton. Rev. Mod. Phys. 35 (1): 1 (1963).

246. E. M. Savitskii and V. V. Baron. Izv. Akad. Nauk SSSR, Ser. Metallurgiya i Gornoe Delo, No. 5:3 (1963).

247. E. Saur. Uber Moglichkeiten zur praktischen Anwendung der Supraleitung Metalls, 16(5):380 (1962). (Russian translation: Ekspress Informatsiya VINITI, Metallovedenie i Termoobrabotka, 30, Ref. No. 113; 14/VIII 1962).

248. B. Kopelman. Materials for Nuclear Reactors, New York, McGraw-Hill Book Company (1959) [Russian translation, Gosatomizdat (1962)].

249. V. V. Pen'kovskii. Effect of Irradiation on Metals and Some Refractory Materials, Acad. Sci. Ukr.SSR Press (1962).

250. J. Kaye and J. Welsch. Collection: Direct Conversion of Heat to Electricity [Russian translation], Atomizdat (1961).

251. Collection: Conversion of Heat and Chemical Energy into Electrical Energy in Rocket Systems [Russian translation edited by A. E. Sheindlin, IL (1963)].

252. H. C. Gatos. The Physics and Chemistry of Ceramics. Edited by C. Klingsberg, Scient. Publ., 1963, p. 196.

253. S. A. Nemnonov, M. F. Sorokina, A. Z. Men'shikov, K. M. Kolobova, and L. D. Finkel'shtein. Fiz. Metal. i Metalloved. 14 (5):535 (1962).

254. J. H. Wernick. The Proceedings of the Conference on Properties of Elemental and Compound Semiconductors, Met. Soc. AIME (1959).

255. J. H. Wernick. Am. Mineralogist 45:591 (1960).

256. N. Shönberg. Acta Chem. Scand. 8 (2):221 (1954).

257. G. Brauer, H. Muller, and G. Kuhner. J. Less Common Metals 4 (6):533 (1962).

258. N. Shonberg. Acta Chem. Scand. 8:225 (1954).

259. H. Lux and E. Procschel. Z. Anorg. Chem. 257:73 (1948).

260. G. Hägg and N. Shönberg. Acta Cryst. 7:351 (1954).

261. Ya. I. Gerasimov, I. A. Vasil'eva, T. P. Chusova, V. A. Geidrikh, and M. A. Timofeeva. Dokl. Akad. Nauk SSSR 134 (6):1350 (1960).

262. G. R. Pierre, W. T. Ebihara, M. J. Poofe, and R. Speiser. Trans. Met. Soc. AIME 224:259 (April 1962).

263. T. A. Babaeva and R. I. Kuznetsova. Collection: Structure of Alloys of Some Systems Containing Uranium and Thorium, Gosatomizdat (1961), p. 423.

264. D. T. Hurd. Introduction to the Chemistry of the Hydrides, New York, John Wiley & Sons, Inc. (1952), [Russian translation, IL (1955)].

265. V. I. Mikheeva. Transition Metal Hydrides, Acad. Sci. USSR Press (1960).

266. F. M. Perel'man. Methods of Representation of Many-Component Systems, Acad. Sci. USSR Press (1959).

267. V. A. Ocheretnyi. Zh. Neorgan. Khim. 5 (7):1588 (1960); Zh. Neorgan. Khim. 6(10):2372 (1961); Two- and Many-Dimensional Sections of Figures and Their Use for Representing Many-Component Systems, Author's abstract of dissertation, Moscow (1963).

268. W. E. Wallace. Annual Review of Physical Chemistry V. 15, USA, California, 1964; Intermetallic Compounds, pp. 109-130.

269. V. A. Bryukhanov, N. N. Delyagin, and R. N. Kuz'min, Zh. Éksperim. i Teor. Fiz. 46:137 (1964).

270. V. I. Nikolaev, Yu. I. Shcherbina, and A. I. Karchevskii. Zh. Éksperim i Teor. Fiz. 44:775 (1963).

271. V. I. Nikolaev, Yu. I. Shcherbina, and S. S. Yakimov. Zh. Éksperim. i Teor. Fiz. 45:1277 (1963).

272. E. Piegger, R. S. Craig. Journ. Chem. Phys. 32: 137 (1963).

273. S. Komura, N. Shikazano. Journ. Phys. Soc. (Japan), 18: 323 (1963).

274. A. M. Bardos and D. I. Bardos, Trans. Met. Soc. AIME, 227: 991 (1963).

275. J. G. Faller, L. P. Scofnick, Trans. Met. Soc. AIME, 227: 687 (1963).

276. I. I. Kornilov. "Development of investigations in the field of metal chemistry," Usp. Khim. 34(1):103(1965).

277. J. C. Wooley, J. H. Phillips, and J. A. Clark, J. Less Common Metals, 6 (6): 461 (1964).

278. I. I. Kornilov, P. B. Budberg, K. I. Shakhova, and S. P. Alisova. Dokl. Akad. Nauk SSSR 161(6): 1378 (1965).

279.   G. V. Samsonov and V. N. Paderno. Izv. Akad. Nauk SSSR; Ser. Met., No. 1:180 (1965).

280.   M. Wells, M. Pickus, K. Kennedy, and V. Zackey, Phys. Review Letters, 12 19:536 (1964).

281.   E. M. Savitskii, V. V. Baron, Yu. V. Efimov. and E. I. Gladyshevskii. Zh. Neorgan. Mat. 1 (2): 208 (1965).

282.   N. Kh. Abrikosov and G. T. Danilova-Dobrayakova. Zh. Neorgan. Mat. 1 (2):204 (1965).